Giant Carpenter Bees:
Taxonomy and Identification of the Species
of the *Xylocopa (Mesotrichia)* Group

Giant Carpenter Bees:
Taxonomy and Identification of the Species of the *Xylocopa (Mesotrichia)* Group

Jonathan Mawdsley

Pineway Press
2017

Copyright © 2017 by Jonathan Mawdsley

All rights reserved. This book or any portion thereof may not be reproduced or used in any manner whatsoever without the express written permission of the publisher except for the use of brief quotations in a book review or scholarly journal.

First Printing: 2017

ISBN-13: 978-1981740970
ISBN-10: 198174097X

Pineway Press
6801 Pineway
University Park, MD, 20782, USA

www.jonathanmawdsley.com

Dedication

This book is dedicated
to Corinne Carter for her help and encouragement,
to James Harrison and Sam Droege
for their good fellowship and companionship in the field,
and to my parents, Ralph and Alice Mawdsley.

Contents

Acknowledgments ix
Abstract xi
Preface xiii
Chapter 1: Introduction 1
Chapter 2: Materials and Methods 5
Chapter 3: African Species 7
Chapter 4: Asian Species 21
References 50
Glossary 53

Acknowledgments

I am greatly indebted to the many colleagues and friends who have graciously made specimens of these bees available to me for study. In particular I wish to thank Seán Brady and Brian Harris of the National Museum of Natural History, Smithsonian Institution (USNM), who very kindly provided me with access to their large collections of these bees as well as the large collection of reprints and files on carpenter bees which had been compiled by the late Paul D. Hurd. Much of the specimen data presented in this book comes from specimens preserved in USNM. For providing me with digital photographs of type specimens in their care, I thank Hadel Go and John Ascher of the American Museum of Natural History (AMNH) as well as David Notton of The Natural History Museum, London (BMNH). For formal permission to study large carpenter bees in the field in the Kruger National Park, Republic of South Africa, I thank Freek Venter and South African National Parks. For their many contributions and gracious assistance with the South African fieldwork, I thank Hendrik Sithole, Guin Zambatis, Patricia Khoza, Adolf Manganyi, Obert Mathebula, Onica Sithole, Velly Ndlovu, Moffat Mambane, Rheinhardt Scholtz, and Thembi Khoza, all of whom work for South African National Parks. James Harrison of the University of the Witwatersrand (formerly of the Transvaal Museum) provided outstanding contributions to these field studies as well as much helpful coordination with various logistical aspects of the research trips to South Africa. Sam Droege of the U. S. Geological Survey's Bee Monitoring Laboratory accompanied me in the field in South Africa in 2016 and provided me with much useful guidance and information about the sampling and collection of bees. Key field volunteers and supporters of this project included Alice, Ralph, and James Mawdsley, as well as Marilyn Shoger. My 2012 field research in South Africa was supported by the H. John Heinz III Center for Science, Economics and the Environment. For sponsoring my trip to the Philippines in 2017, I thank Ron Regan, Mark Humpert, and the Association of Fish and Wildlife Agencies.

Abstract

Keys and photographic illustrations are presented for the identification of males and females of the twelve valid species in the African and Asian "Mesotrichia Group" within the genus *Xylocopa* Latreille, including the seven African species of subgenus *Mesotrichia* Westwood, the three Asian species of subgenus *Platynopoda* Westwood, the single species of subgenus *Hoplitocopa* Hurd and Moure, and the single species of subgenus *Hoploxylocopa* Hurd and Moure.

In the subgenus *Xylocopa (Mesotrichia)*, seven valid species are recognized: *X. (M.) chapini* (LeVeque); *X. (M.) combusta* Smith; *X. (M.) flavorufa* (DeGeer); *X. (M.) ignescens* (LeVeque); *X. (M.) mixta* Radozkowski; *X. (M.). subcombusta* (LeVeque); and *X. (M.) torrida* (Westwood). The holotype of *Mesotrichia perpunctata* LeVeque (which was described from female specimens only) represents the female of *Mesotrichia ignescens* LeVeque (which was described from male specimens only); the name *X. (M.) ignescens* has page priority and thus *X. (M.) perpunctata* must be placed in synonymy with *X. (M.) ignescens*, new synonymy. *Xylocopa (M.) fervens* Lepeletier was based on material from the Cape of Good Hope which had been erroneously labeled as having been collected in "Cayenne" and represents a synonym of the southern African species *X. (M.) flavorufa*. *Mesotrichia lautipennis* Cockerell is placed in synonymy with *X. (M.) flavorufa* based on study of the unique holotype specimen which is a teneral individual of *X. (M.) flavorufa*. Three taxa originally described as subspecific or infrasubspecific entities are treated here as synonyms of the nominate forms: *X. (M.) flavorufa* variety *harrarensis* Vachal and *X. (M.) flavorufa* variety *kristenseni* Friese are treated as synonyms of *X. (M.) flavorufa*, while *X. (M.) torrida* variety *gramineipennis* Friese is treated as a synonym of *X. (M.) torrida*.

In the subgenus *Xylocopa (Platynopoda)*, three valid species are recognized: *X. (P.) latipes* (Drury); *X. (P.) perforator* Smith; and *X. (P.) tenuiscapa* Westwood. *Xylocopa latipes* var. *magnifica* Cockerell is placed in synonymy with *X. tenuiscapa* Westwood. *Xylocopa marginella* Lepeletier is placed as *species incertae sedis* pending rediscovery of the type specimen(s), while *X. (P.) yunnanensis* Wu is provisionally transferred to the subgenus *Biluna* Maa of the genus *Xylocopa*.

A single species is recognized in each of the subgenera *Xylocopa (Hoplitocopa)* and *Xylocopa (Hoploxylocopa)*: *Xylocopa (Hoplitocopa) assimilis* Ritsema and *Xylocopa (Hoploxylocopa) acutipennis* Smith. *Mesotrichia kerri* Cockerell is placed in synonymy with *Xylocopa acutipennis* Smith.

Preface

This book is part of a series of publications which are intended to facilitate the identification of species in certain poorly-known groups of the genus *Xylocopa* Latreille, or large carpenter bees. Bees in this genus are tremendously important as pollinators of flowering plants, both in agricultural settings and in natural ecosystems. Unfortunately, the taxonomy of the genus *Xylocopa* is extremely difficult, and there are few reliable identification guides available for many of the most common and abundant groups of these bees. This guidebook brings together information on the identification of species in the so-called "Mesotrichia Group," a distinctive group of bees found throughout sub-Saharan Africa and tropical and subtropical Asia.

The twelve species of bees included in this book are some of the largest and most beautiful bees on earth. I have been very fortunate to have had the opportunity over the last decade to study and photograph species of these bees in the field, primarily in South Africa's Kruger National Park during my annual research trips between 2006 and 2017, and also in the Philippine Islands during a visit in 2017.

In addition to their significance as pollinators, preserved adult specimens of these bees often turn up in the global insect specimen trade, where they are used in art projects on account of the beautiful iridescent colors on the bees' wings. It is not clear whether the harvest of these bees for the global specimen trade is actually a sustainable use of these large and beautiful insects. However, many species of carpenter bees – including some species of the "Mesotrichia Group" can actually become minor pests around human habitations, due to their habit of excavating their nests in the exposed wood portions of human structures. The harvesting of adult bees for the specimen trade, if done sustainably, could potentially be a helpful strategy for reducing the damage caused by these bees by their nesting activities.

Large carpenter bees often have close associations with human beings, excavating their nests in human structures and visiting flowers in gardens, agricultural areas, and other sites near human habitations. The adult bees are often highly conspicuous when visiting flowers, establishing and defending nesting territories, or excavating their nests. Species of these bees often come to the attention of naturalists, insect collectors, and even the general public. It is my hope that the publication of this guidebook will stimulate other entomologists, biologists, and naturalists to conduct further studies of the biology, life history, and pollination ecology of these majestic insects.

Chapter 1: Introduction

This book provides keys and illustrations to help identify some of the largest and most spectacular bees on earth: the twelve species of the "Mesotrichia Group" within the larger genus *Xylocopa* Latreille. The bees included in this guide were all classified by Michener (2007) as members of the subgenus *Mesotrichia* Westwood within the genus *Xylocopa*. Species of the genus *Xylocopa* are known generally as "large carpenter bees" due to their large body size and the fact that females excavate nests in wood (Gerling et al. 1989). The species in the "Mesotrichia Group" are so enormous in size that they truly merit the common name of "giant carpenter bees," a name which has been given by southern African entomologists to certain African members of this group.

The genus *Xylocopa* is a "mega-diverse" lineage of bees, containing approximately 700 described species (Hurd and Moure 1963; Michener 2007). Both the males and the females of *Xylocopa* species visit flowers (Figures 1, 2). Many of these bees are important pollinators of flowering plants, in agricultural systems as well as natural ecosystems (Keasar 2010; Mawdsley et al. 2016). Information about flower visitation and pollination by species in the "Mesotrichia Group" is limited, although Raju and Rao (2006) studied flower visitation by species in India, and Watmough (1974), Eardley (1983), and Mawdsley et al. (2016) studied the biology and floral associations of species in southern Africa. Much important work remains to be done in order to identify the floral associates of these large and colorful bees, and to determine the significance of these giant carpenter bees as possible pollinators of crop species and other flowering plants.

Many species of large carpenter bees excavate nests within dry wood, including standing dead trees and dead tree branches, and many of these species will also excavate nests within human structures made out of wood (Hurd and Moure 1963; Gerling et al. 1989; Michener 2007). While relatively little has been published to date about the nesting biology of the species in the "Mesotrichia Group," the very large body size of these bees and the relative abundance of certain species suggest that these bees could certainly have the potential to become minor pests if they excavated their nests within wooden human structures.

This taxonomic treatment and guidebook includes all of the species recognized by Michener (2007) as belonging to the subgenus *Mesotrichia* of the genus *Xylocopa*. In an earlier publication (Mawdsley 2015), I noted that the African and Asian species of this group different significantly in their external morphological characteristics, and recommended reverting to an older classification of these bees which had been published by Hurd and Moure (1963). These authors divided the species in Michener's subgenus *Mesotrichia* into four morphologically distinct subgenera: *Hoplitocopa* Hurd and Moure, *Hoploxylocopa* Hurd and Moure, *Mesotrichia* Westwood, and *Platynopoda* Westwood. I recognize all four of these subgenera here, although it is likely that at least two (and possibly more) of these subgenera actually represent derived forms within other Old World carpenter bee lineages, particularly the highly diverse and poorly-known subgenus *Koptortosoma* Gribodo. Final determination of the validity of these four subgeneric names will require the completion and publication of more comprehensive phylogenetic studies of the species of the genus *Xylocopa*.

Females of species of *Xylocopa* are defended by powerful stings, and the sting apparatus of female bees in the "Mesotrichia Group" is among the largest of all bees. This formidable defense apparently serves as a deterrent to some (but not all) vertebrate predators, and as a result many other species of bees and even other insects such as flies have evolved body shapes and color patterns that resemble those of the species in this group. In southern Asia, several smaller-bodied carpenter bees of the Asian subgenus *Xylocopa (Biluna)* Maa strongly resemble certain Asian species in the "Mesotrichia Group" in their overall appearance and wing iridescence. These species of subgenus Biluna are frequently confused with species of the "Mesotrichia Group," both in major insect collections and also in the global insect specimen trade. Females of the two groups can be readily separated by the structure of the scutellum, which is broadly rounded in the species of subgenus *Biluna*, but has a strong transverse ridge or carina in species of the "Mesotrichia Group." To aid in identification of these other bees, I have also included illustrations of the two species of the subgenus *Biluna* which are most commonly confused with species of the "Mesotrichia Group" at the end pof this guidebook. Because females of the species of the subgenus *Biluna* are also defended by powerful stings, the overall resemblances between these bee species are likely due to convergence on a common or shared color pattern through natural selection by predators such as birds, a process which is known as Müllerian mimicry (Wickler 1968).

Because this guidebook is intended for practical use by a more general audience of general biologists, naturalists, bee enthusiasts, and entomologists, I have divided the species of the "Mesotrichia Group" among two chapters: one chapter for the African species, the second chapter for the Asian species. For each fauna, there is a key to species, followed by individual species accounts which provide a summary of taxonomic information, a brief diagnosis based on external morphological features of males and females, any notes, and a detailed list of specimens that I examined in preparing this guide. Color photographic illustrations and country-level distribution maps are provided for each species at the end of the guide.

Giant Carpenter Bees

Chapter 2: Materials and Methods

This guidebook covers the taxonomy and identification of all of the species of the genus *Xylocopa* which were included by Michener (2007) in the subgenus *Mesotrichia* Westwood. The taxonomic portions of this work are based on studies of dead, pinned, preserved museum specimens of these bees, including numerous specimens from the very large collection of carpenter bees which was assembled by Paul Hurd, the former curator of bees at the Department of Entomology, National Museum of Natural History, Smithsonian Institution (USNM). This collection includes holotype, paratype or syntype specimens of species which were described by many of the major taxonomists who studied the taxonomy of these species of genus *Xylocopa*, including T. D. A. Cockerell, T. C. Maa, N. LeVeque, and H. Friese. The collection also includes numerous specimens which were authoritatively identified by comparison with known type specimens by P. D. Hurd and M. A. Lieftinck, among other workers. It also includes numerous specimens of carpenter bees which were reared by Hurd and his colleagues and contemporaries from nests, an approach which can help facilitate the accurate association of the males and females of the individual species (Hurd and Moure 1963). I also received specimens of these bees from a number of private collectors and individuals. I particularly thank Brent Karner of BioQuip Bugs, Inc., for his assistance in providing specimens for this study. In addition to the physical specimens, I also examined digital images of type specimens of *Xylocopa* species in the American Museum of Natural History, New York, and The Natural History Museum, London.

The morphological nomenclature that is used in this work follows that published in the classic study of carpenter bee morphology and taxonomy by Hurd and Moure (1963). A brief glossary is provided at the end of this guidebook covering those terms that may not be familiar to a general biologist or an entomologist without specialized training in hymenopteran morphology and systematics. Definitions of many of the other standard terms that are used in insect morphology can be found at the Wikipedia page "Glossary of Entomology Terms" < https://en.wikipedia.org/wiki/Glossary_of_entomology_terms>

    I have been fortunate to have had the opportunity to study species of these bees in the field, in both the Philippines and the Republic of South Africa. My studies in South Africa are part of a longer study, now in its twelfth year, which is investigating the biology, life history, and conservation status of insect pollinators in the Kruger National Park. The publication by Mawdsley et al. (2016) provides an overview of this project, a description of our basic research approach and methods, as well as a synopsis of the key research findings and identification materials and photographs of major pollinator taxa within the Kruger National Park.

    While this manuscript was in the process of development, I had the opportunity to visit the Philippine Islands in order to attend the twelfth Conference of the Parties of the United Nations Convention on Migratory Species as a representative of the Association of Fish and Wildlife Agencies. During my stay in the Philippines, I had the opportunity to observe and photograph adults of the species *Xylocopa latipes*.

Chapter 3: African Species

All of the African species of the "Mesotrichia Group" belong to the subgenus *Mesotrichia* Westwood proper, as defined by Hurd and Moure (1963). The type species of the genus *Mesotrichia* is *Mesotrichia torrida* Westwood, by original monotypy.

African species of this group may be recognized by the following combination of characteristics (Mawdsley 2017): **Female:** Body size very large; scutellum with a large transverse ridge or carina, carina extending back over the base of the mesosoma; integument black, with black, red, orange-yellow, or brown pubescence. **Male:** large-bodied, with middle pair of legs strongly modified and bearing spines at base, elongate setae, and/or distinct patches of pubescence.

The taxonomy of most of the African species which are currently recognized in this group was reviewed by LeVeque (1928) in a paper which was based largely on specimens in the collections of the the American Museum of Natural History. Subsequent to this publication, many of the species included by LeVeque (1928) in the subgenus *Mesotrichia* were transferred by Hurd and Moure (1963) to the subgenus *Koptortosoma* Gribodo. Species of the two subgenera are similar, but females of *Koptortosoma* species generally are smaller-bodied and have a smaller and less prominent transverse ridge on the scutellum, while males of *Koptortosoma* are smaller-bodied and lack the distinctive modifications to the middle pair of legs found in species of subgenus *Mesotrichia*.

Key to African Species of the "Mesotrichia Group"

1) Eyes massive, covering at least half of head capsule, sting apparatus absent, males... 2
- Eyes smaller, covering less than half of head capsule, sting apparatus present, females... 8

Giant Carpenter Bees

2) Apex of abdomen with pale yellow or orange pubescence.......... 3
- Apex of abdomen with black pubescence............................. 7
3) Mesosoma with yellowish-brown, red, or orange pubescence as well as black pubescence... 4
- Mesosoma with black pubescence only................................. 6
4) Wings with predominantly bluish-green iridescence, sometimes with violet highlights.................................... *X. chapini*
- Wings with uniformly violet or bluish-violet iridescence............. 5
5) Frons with black and orange-red pubescence............ *X. flavorufa*
- Frons with orange-red pubescence only....................... *X. mixta*
6) Wings with bluish-green iridescence..................... *X. combusta*
- Wings with violet iridescence........................... *X. subcombusta*
7) Wings with bluish-green iridescence........................ *X. torrida*
- Wings with violet iridescence, sometimes also with bluish-green iridescence.. *X. ignescens*
8) Pubescence of mesosoma entirely black............................. 9
- Pubescence of mesosoma yellowish-brown, red, or orange, at least in part.. 10
9) Wing iridescence uniformly brilliant greenish-blue.... *X. combusta*
- Wing iridescence mostly or entirely violet, sometimes with greenish-blue at base and apex........................ *X. subcombusta*
10) Pubescence of mesosoma red or orange........................... 11
- Pubescence of mesosoma brown or yellowish-brown............... 12
11) Frons with black and orange-red pubescence........... *X. flavorufa*
- Frons with orange-red pubescence only....................... *X. mixta*
12) Apex of abdomen black.. *X. torrida*
- Apex of abdomen paler in color.. 13
13) Wings with violet iridescence; yellow pubescence limited to terminal segment of metasoma......................... *X. ignescens*
- Wings with bluish-green iridescence, sometimes with violet iridescent highlights; yellow pubescence on apical two segments of metasoma... *X. chapini*

Xylocopa (Mesotrichia) chapini
(LeVeque 1928:5)

Figure 3: a: female PARATYPE, dorsal view; **b:** male PARATYPE, dorsal view; **c:** female, front of head; **d:** female, metasoma, showing apical setae; **e:** distribution map, by country.

Mesotrichia chapini LeVeque (1928:5), type locality "Faradje, Congo," type material examined: PARATYPES, Faradje, Congo, 25 40 E, 3 40 N, XI.1912, Lang and Chapin collectors (5 females and 3 males); PARATYPES, same data but XII.1912, Lang and Chapin collectors (3 females and 1 male), PARATYPE, same data but IV.1911, Lang and Chapin collectors (1 male); PARATYPE, Garamba, Congo, 29 40 E, 4 10 N, Lang and Chapin collectors (1 female) (all USNM).

Brief Diagnosis: Female: Overall body length 19-25 mm, metasomal width 11-13 mm; integument black; vesture black except for yellowish-brown pubescence on dorsal and lateral surfaces of mesosoma and a transverse band of yellowish-brown pubescence on the apical tergite of metasoma; wings dark brownish-black with bluish-green iridescence, medially with feeble violet iridescence. **Male:** Overall body length 22-26 mm, metasomal width 11-13 mm; integument black, clypeus and scape yellow; vesture black, mesosoma with yellowish-brown pubescence on dorsal and lateral surfaces, two apical tergites of metasoma with yellowish-brown pubescence; wings dark brownish-black, with uniform bluish-green iridescence, occasionally with violet iridescent highlights.

Notes: Specimens from the type series of this species are very similar to and may ultimately prove through further study to be a regional subspecies or variant of the common and widespread species *X. flavorufa*. However I think it is best at the present time to maintain these two forms as separate species, based on differences in the wing iridescence, coloration of the dorsal pubescence, and the male reproductive structures, which were illustrated by LeVeque (1928).

Xylocopa (Mesotrichia) combusta
Smith (1854:350)

Figure 4: a: female, dorsal view; **b:** male, dorsal view; **c:** female, metasoma, showing apical setae; **d:** distribution map, by country.

Xylocopa combusta Smith (1854:350), type locality "Congo," species concept based on the two specimens listed below from Leopoldville and Matadi in the Democratic Republic of the Congo, which were identified and labeled by N. LeVeque as belonging to these species, based on her own comparisons of these specimens with primary type material in the collection of T. D. A. Cockerell (as noted by LeVeque 1927:7).

Xylocopa taczanovskii Radoszkowski (1876:350), synonymy by Vachal (1900:108). Type locality "d'Abyssinie" (= Ethiopia; Radoszkowski 1876:350).

Brief Diagnosis: Female: Overall body length 25-27 mm, metasomal width 12-13 mm; integument black; vestiture black, except for a small well-defined patch of bright red pubescence/setae at the apex of the metasoma; wings dark brownish-black with brilliant blue-green iridescence, not violet. **Male:** Overall body length 21-30 mm, metasomal width 9-13 mm; integument black, except for yellow markings on clypeus and antennae; vestiture black; wings black with brilliant blue-green iridescence.

Additional Materials Examined: ANGOLA: Bom Jesus, 22.VIII.1949, B. Malkin (1 female); Chitau, 1-12.VIII.1925 (1 male); Luanda, 18.VII.1957, E. S. Ross and H. E. Leech, collectors (1 male), same locality but VI-VII.1957-1958 (1 male), same locality but elevation 50 m, 3.VI.1958 (1 female), same locality (1 male); Saurimo, XI.1949, P. Eduardo (1 male). DEMOCRATIC REPUBLIC OF THE CONGO [Belgian Congo]: Leopoldville, 4 25 S, 15 20 E, 2-11.VI.1909, Lang and Chapin collectors (1 female); Matadi, 22.VII.1957, E. S. Ross and H. E. Leech, collectors (1 male). REPUBLIC OF THE CONGO [French Equatorial Africa]: Point Noire, 12-13.VI.1957 (3 males). Eardley (1987) also records this species from: Equatorial Guinea, Mozambique, Sierra Leone, Somalia, and Tanzania.

Xylocopa (Mesotrichia) flavorufa
(DeGeer 1778:605)

Figure 1: females visiting flowers of *Karomia speciosa* (Hutchinson & Corbishley) R. Fernandes (Lamiaceae), photographed November 26, 2017, by the author at Skukuza, Mpumalanga, Republic of South Africa, in the Kruger National Park.

Figure 5: a: female, compared with type by P. D. Hurd, with extensive red pubescence on mesosoma, dorsal view; **b:** female, with reduced red pubescence on mesosoma, dorsal view; **c:** male, dorsal view; **d:** female, front of head; **e:** female, metasoma, showing apical setae; **f:** distribution map, by country.

Apis flavo-rufa DeGeer (1778:605), type locality "Cap-du-bonne-esperance," species concept based on female specimen which was compared with the holotype in the Stockholm Museum by P. D. Hurd and labeled "COMPARED WITH TYPE STOCKHOLM PD HURD 1964," collected at Port Elizabeth, Cape Province 20.II.1930, J. S. Taylor, ex nest in *Melia azedarach* L. (USNM).

Xylocopa fervens Lepeletier (1841:196), new synonym, type locality given as "Cayenne" which was apparently an error for "Cape of Good Hope" (Smith 1874:293).

Xylocopa flavorufa variety *harrarensis* Vachal (1922:988), type locality "Harar," species concept based on three specimens: Ethiopia: Addis Ababa, 5.VII.1912 (1 female), same data but 28.VII.1920 (2 females), same data but 21.VIII.1920 (1 female), identified as *X. (M.) flavorufa* var. *harrarensis* by P. D. Hurd.

Xylocopa flavorufa variety *kristenseni* Friese (1911:685), type locality "Harar" and "Br. Abessinien, Südost-Gebiet," species concept based on three specimens: Ethiopia: Addis Ababa, 16.VII.1920 (1 male), 11.IX.1920 (1 male), 22.VI.1958, R. E. Fontaine (1 male), identified as *X. (M.) flavorufa* var. *kristenseni* by P. D. Hurd.

Mesotrichia lautipennis Cockerell (1933:457-458), new synonym, type locality "Temke, Congo," type material examined: HOLOTYPE, female, "Belgian Congo, Katanga, Temke, 30.VII-9.VIII.1931" (BMNH).

Brief Diagnosis: Female: Overall body length 21-28 mm, metasomal width 9-13 mm; integument black; frons with mixed black, white, and yellow or red pubescence, black pubescence always present; mesosoma and apex of metasoma with dense red, yellow, or orange pubescence; remaining vestiture black; wings dark brownish-black with brilliant violet iridescence. **Male:** Overall body length 22-27 mm; metasomal width 10-12 mm; frons with black pubescence, at least in part; mesosoma and apex of metasoma with dense red, yellow, or orange pubescence; wings dark brownish-black with brilliant violet iridescence.

Notes: Both sexes are similar to and can be confused with those of *X. mixta*. The females of these species can be readily separated by the presence of black pubescence (at least in part) on the frons of *X. flavorufa*, which is lacking entirely in *X. mixta*. Males of these two species can be separated reliably only by structures of the male genitalia (Eardley 1983), although males of *X. mixta* have dense orange, yellow, or red pubescence on the vertex of the head which is usually also mixed with black pubescence in males of *X. flavorufa*.

Lepeletier (1841:196) expressed doubts about the correctness of the locality label in his original description of *X. fervens* and actually stated that "Je croiras plutôt cette espèce de Bonne-Espérance." Smith (1874:293) further states "St. Fargeau [Lepeletier] gives the above locality [Cayenne] but I am inclined to think that he is wrong, and that the insect is a fine variety of *X. flavo-rufa* from S. Africa." I concur with Smith's assessment and accordingly place *X. fervens* in synonymy with *X. flavorufa*.

David Cotton of The Natural History Museum, London, very kindly sent me a set of high-resolution digital photographs of the unique female holotype of *Mesotrichia lautipennis* Cockerell, which is a teneral female of *X. flavorufa*.

The two varieties of this species described from Ethiopia represent normal color variants within the species and do not appear to represent geographical entities which could be named as subspecies. I therefore treat these names as synonyms of *X. flavorufa*.

Additional Material Examined: ANGOLA: Benguella, Ganda, IV.1953, W. Hellmich (1 female); Chitau, 1-12.VIII.1925 (1 female); "Angola" (2 females).   DEMOCRATIC REPUBLIC OF THE CONGO [Belgian Congo]: Katanga, Kabongo, 28.X.1953, C. Seydel (1 male), Lwiro River, 47 km N Bukavu, 2000 m, 1.IX.1957, E. S.

Ross and R. E. Leech (1 female), same data except 2100 m, 27.VIII.1957 (1 female), same data except 29.VIII.1957 (1 female). ETHIOPIA: Awasa, VII.1972 (1 male); 15 km N of Comboichia, 6500 feet, 15.I.1960, E. S. Ross (1 female); Dedo, 17.XI.1974, J. Goodyear (1 female). KENYA: Kaimosi, Kahamega, 19-25.II.?, A. Loveridge (5 females); Kisii, VII, A. Loveridge (2 females); Kwale, 400 m, 5.XI.1957, E. S. Ross and R. E. Leech (4 females); Lake Naivasha shore, 6230 feet, 1948, N. A. Weber (1 female); Mombasa, 28.XI.1952, Lindemann and Pavlitzki (1 female); Nairobi, 8.V.1915, A. Loveridge (1 male), same data but 1915 (1 female); same data but 1916 (1 female); Nairobi, Kavuvu, 14.VIII.1919, A. Loveridge (1 female); Tana River, Chanler, 1892-1893 (1 female). MOZAMBIQUE: 27 miles E. of Vila Manica, 700 m, 13.III.1958, E. S. Ross and R. E. Leech (1 female). REPUBLIC OF SOUTH AFRICA: Cape Province: Algoa Bay, Dr. Brauns (1 female); Mossel Bay, IV.1921, R. E. Turner (1 female); Port Elizabeth, 24.II.1930, J. S. Taylor, ex nest in Melia azedarach (1 female). Natal: Benvie, near Karkloof, 17.X.1966, at flowers of Azalea (5 males, 3 females); Hilton, 30.I.1906, J. S. Taylor (1 female), same data but 10.II.1906 (1 female); Karkloof, 6.I.1967, J. S. Taylor (1 male), same data except 6.II.1967 (1 male); Umkomaas, VII.1948, A. L. Capener (1 female). TANZANIA [including labels from localities in Tanganyika Territory and Zanzibar]: 15 miles S of Handeni, 13.XI.1957, 630 m, E. S. Ross and R. B. Leech (1 male, 1 female); Kilimandjaro, Marangu, 1500 m, 30.X.1958, C. Lindemann (1 female); Kilimanjaro, W. L. Abbott (4 females); Mamba, SE side Kilimanjaro, 1440 m, 29.X.1957, E. S. Ross and R. E. Leech (1 female); Marienhof, Ukerewe, 1911, Conrads (1 female); Mgeta, Uluguru Mountains, 800 m, 15.XI.1957, E. S. Ross and R. E. Leech (1 female); Morogoro, 12.II.?, A. Loveridge (1 male), same data but 18.IV.? (1 male); Songea, Litembo, 28.VIII.1952, Lindemann and Pavlitzki (1 male), same data but 18.XI.1958, Lindemann (1 female); Tabora, 29.X.1956, W. E. Kerr (1 male); Tanga, 13.VII.1952, Lindemann and Pavlitzki (1 female), Ukerewe L., F. Conrad (1 female); Usambara, VI.1903, P. Herbst (1 female); Zanzibar, C. Cooke (1 female). UGANDA: Between Kafu R. and Kigema, Hoima-Kampala Road, 3600-3800 feet, 1-3.I.1912, S. A. Neave (1 female); Kampala, 8.VI.1963, J. Owen (1 female), same data but 30.VI.1963 (1 female), same data but 9.VII.1963 (1 male, 5 females), same data but 12.VII.1963 (1 female), same data but 13.IX.1963 (1 female), same data but 26.X.1963 (1 male), same data but 1963 (4 females). ZAMBIA [Northern Rhodesia]: Lusaka, 11.IX.1980, Ridgeway, pollen on pronotum (1 female); Senga Hill, 40 miles N of Abercorn, 1580 m, 11.XII.1958, E. S. Ross and R. E. Leech (1 female). ZIMBABWE [Rhodesia]: Bulawayo, 23.XI.1924, S. H. Stevenson (1 female); Salisbury, 7.XI.1951, Zumpl (1 female), 24.IV.1965, R. H. Watmough (1 male, 3 females), same data but 6.V.1965 (1 female), same data but 8.V.1965 (1 female), same data except 14.V.1965 (3 females), same data except 17.XII.1965 (1 female); Salisbury, 24 Chester Road, Avondale, died 8.X.1964, R. H. Watmough (1 female); Salisbury, died in Avondale observation area, 5.II.1965, R. H. Watmough (1 male), same data except 5.III.1965 (1 male), same data except 10.III.1965 (1 female), same data except 4.VI.1965 (2 males, 1 female), same data except 8.VI.1965 (1 male), same data except 23.VI.1965 (1 female), same data except 12.XII.1965 (1 male), same data except 1965 (2 males, 2 females); Warren Hills, 4.V.1965, ex mite experiment, R. H. Watmough (2 females), Warren Hills, 10.IV.1965, R. H. Watmough (1 female), same data but 24.IV.1965, (4 males, 3 females), same data but 4.V.1965 (1 female). "Africa" (1 male, 2 females). Eardley (1987) also records this species from: Burundi, Cameroon, Malawi, Namibia, Nigeria, Rwanda, and Swaziland.

Xylocopa (Mesotrichia) ignescens
(LeVeque 1928:6)

Figure 6: a: female, compared with type, dorsal view; **b:** male, compared with type, dorsal view; **c:** female, metasoma, showing apical setae; **d:** distribution map, by country.

Mesotrichia ignescens LeVeque (1928:6), type locality "Banana, Congo," type material examined: HOLOTYPE, male, "Banana, Congo, 6o S 12o 20' E, VIII,1915, Lang & Chapin, Collectors" (AMNH).

Mesotrichia perpunctata LeVeque (1928:10), new synonym, type locality "Boma, Congo," type material examined: HOLOTYPE, female, "Boma, Congo, 13o 0' E 6o 0' S, VI.16.1915, Lang & Chapin,

Collectors" (AMNH); PARATYPE, female, "Malela, Congo, 6o S, 12o 40' E, 5.VII.1915, Lang & Chapin, Collectors" (USNM).

Brief Diagnosis: Female: Overall body length 23-27 mm, metasomal width 12-13; integument and vestiture black; wings dark brownish-black with deep violet iridescence across most of surface. **Male:** Overall body length 23-29 mm, metasomal width 10-12 mm; integument black, clypeus and antennae with yellow or yellowish-brown markings; vestiture black, except for pale brown pubescence/setae on pronotum, thoracic pleurae, protibiae, protarsi, mesotibiae, mesotarsi, metatibiae, metatarsi, and apex of abdomen; wings dark brownish-black with brilliant greenish-blue iridescence.

Notes: LeVeque (1928) described the male and female of this species under separate names in the same publication. The name *X. ignescens* has page priority and thus is accepted here as the correct, valid name for the species. Both sexes of this species are similar to those of *X. torrida* but the wings in *X. ignescens* are dark with extensive areas of deep violet iridescence in both sexes (sometimes mixed with bluish-green in the male). I thank Hadel Go and John Ascher of the American Museum of Natural History who very kindly provided me with high-resolution digital images of the holotypes of these two taxa.

Additional Material Examined: CAMEROON: Colaris, Sidamo, 3.IV.1953 (2 males), Ebolowa, 8.XII.1931 (1 male). GABON: N'Djole (1 male). GHANA: Kade, University of Ghana Agricultural Research Station, 11.IX.1965 (1 male). NIGER: Crosse (1 male). Nigeria: Sapoba, VII.1966 (1 male). UGANDA: Masindi, Budongo F. 31.VIII.1970, D. H. Messersmith (1 female).

Xylocopa (Mesotrichia) mixta
Radozkowski (1881:199)

Figure 7: a: female, with reduced red pubescence on mesosoma, dorsal view; **b:** female, with extensive red pubescence on mesosoma, dorsal view; **c:** male, dorsal view; **d:** female, front of head; **e:** female, metasoma, showing apical setae; **f:** distribution map, by country.

Xylocopa mixta Radozkowski (1881:199), type locality "Huilla (Anchieta)," type material: According to Eardley (1983:24) the type material of this species could not be traced. My concept of this species is accordingly based on the descriptions and illustrations provided by Eardley (1983:24). According to Hurd and Moure (1963:310) the type locality is in Angola.

Brief Diagnosis: Female: Overall body length 23-29 mm, metasomal width 11-13 mm; integument black; vestiture of frons, vertex of head, protibiae; mesosoma, and apex of metasoma bright orange-red to red; vestiture of head lacking black pubescence; wings dark brownish-black with brilliant violet iridescence. **Male:** Overall body length 19-31 mm, metasomal width 11-12 mm; integument black; vestiture of head, mesotibiae and metatibiae, mesosoma, and apex of metasoma bright orange-red to red; wings dark brownish-black with brilliant violet iridescence.

Notes: Both sexes can be confused with those of *X. flavorufa* but can be readily separated in most cases by the coloration of the pubescence on the frons, which is entirely reddish-orange in *X. mixta* but reddish-orange and black in *X. flavorufa*.

Additional Material Examined: ANGOLA: Benguela (1 female); Chitau, 1-12.VIII.1925 (1 male, 8 females); no locality specified (2 males). DEMOCRATIC REPUBLIC OF THE CONGO [Belgian Congo]: Lwiro River, 17 km N Bukavu, 29.VIII.1957, 2100 m, E. S. Ross and R. E. Leech (1 male). KENYA: Nairobi, 18.XII.1915, A. Loveridge (1 male), same data except 1915 (1 male). Malawi [Nyasaland]: 13 miles SE of Fort Hill, 1310 m, 20.XI.1958, E. S. Ross and R. E. Leech (1 female). TANZANIA [Tanganyika Territory and Zanzibar]: Kilossa, 22.I.1922, A. Loveridge (1 male), same data except 23.I.1922 (1 male); Kigonsera, 1903 (1 male), Marienhof, Ukerewe, 1911, Conrads (1 male); Morogoro, 6.IV.?, A. Loveridge (1 male); Nguelo, Usambara, H. Rolle (1 male); Zanzibar, C. Cooke (1 male). ZAMBIA [Northern Rhodesia]: Samfya, L. Bangweulo, 15.I.1959, visiting Spheostylus sp., R. H. Watmough (1 female); no locality specified, II.1953 (1 male). ZIMBABWE [Rhodesia]: Bulawayo, V.1916 (1 male). "Africa" (1 male).

Xylocopa (Mesotrichia) subcombusta
(LeVeque 1928:9)

Figure 8: a: female PARATYPE, dorsal view; **b:** male, PARATYPE, dorsal view; **c:** female, metasoma, showing apical setae; **d:** distribution map, by country.

Mesotrichia subcombusta LeVeque (1928:9), type locality "Banana, Congo," type material examined: HOLOTYPE, female, Banana, Congo, 6 S, 12 20 E, VIII 1915, Lang and Chapin collectors. PARATYPES, 1 female and 2 males, same data as holotype; PARATYPE, male, Boma, Congo, 13 0 S, 6 0 E, 18.VI.1915, Lang and Chapin collectors; PARATYPE, female, Malela, Congo, 6 S, 12 40 E, 8.VII.1915, Lang and Chapin collectors; PARATYPE, female, Zambi, Congo, 6 S, 12 50 E, 22.VI.1915, Lang and Chapin collectors; PARATYPE, female, same data (all USNM).

Brief Diagnosis: Female: Overall body length 20-27 mm, metasomal width 10-13 mm; integument black, vestiture black except for a small area of bright red pubescence at the apex of the metasoma, wings dark brownish-black with brilliant deep violet iridescence, sometimes with brighter greenish-blue iridescence also at base and apices. **Male:** Overall body length 22-28 mm, metasomal width 11-14 mmm; integument black, vestiture black; clypeus and antennae with yellow markings; wings dark brownish-black with deep violet iridescence.

Notes: Both sexes are similar in appearance to those of *X. combusta* but the wings of *X. subcombusta* have mostly violet iridescence while the wings of *X. combusta* have bluish-green iridescence.

Additional Material Examined: ANGOLA: Dunda, Luanda, 21.XI.1949, B. Malkin (1 female); Luanda, VI and VII, 1957 and 1958 (1 female); Malange, 11.IX.1949, Malkin (4 females); Quitondo, Dist. Calulo, 11.VIII-2.IX.1967; Saurimo, XI.1949, P. Eduardo (3 females). "Congo": 30.XII.? (2 females), 1.II.? (1 female), no further data (1 female). DEMOCRATIC REPUBLIC OF THE CONGO [Belgian Congo]: Kabinda, 6.8 S, 4.21 E (1 female); Katanga, Kabonga, 27.X.1953, C. Seydel (1 male), same data except 28.X.1953 (1 male); Matadi, 22.VII.1957, E. S. Ross and S. B. Leech (2 females). ETHIOPIA: Awaso, 24.VI.1971 (1 female); Nazareth, 12.VI.1958, P.

Jolivet (5 females); Sidamo, Tullo-Awaso, 4.VII.1971 (1 female). NIGERIA: Jos, 14.III.1949, B. Malkin (1 female), same locality but 29.X.1961, nesting in structural wood, E. J. Gerberg (1 female); Olokemeji, Ibadan (11 females); North Region, Panyam Fish Farm, 8-16.IV.1967, J. C. Geest (1 female). REPUBLIC OF THE CONGO [French Equatorial Africa]: Point Noire, 12-13.VI.1957 (1 female); no locality specified, XII.1978 (1 male).

Xylocopa (Mesotrichia) torrida
(Westwood 1838:113)

Figure 9: a: female, dorsal view; **b:** male, dorsal view; **c:** female, metasoma, showing apical setae; **d:** distribution map, by country.

Mesotrichia torrida Westwood (1838:113), type locality "in Africa tropicali occidentali," type material: According to Eardley (1983:25) the type material of this species (consisting of an unspecified number of male specimens from an unknown locality or localities in tropical Africa) could not be traced. My concept of this species is accordingly based on the treatment and illustrations provided by Eardley (1983:25).

Xylocopa crassa Lepeletier (1841:204), synonymy by Smith (1874:260), type locality unknown; "sans indication de Patrie" (Lepeletier 1841:204).

Xylocopa torrida gramineipennis Friese (1922:7), type locality "Goldküste, Westafrika," material examined: Republic of the Congo [French Equatorial Africa]: Point Noire (1 female), identified as *X. t. gramineipennis* by P. D. Hurd.

Brief Diagnosis: Female: Overall body length 24-28 mm, metasomal width 9-13 mm; integument black, vestiture entirely or almost entirely black with at most a few red or brown setae at the apex of the metasoma; wings black, with brilliant greenish-blue iridescence. **Male:** Overall body length 24 mm, metasomal width 13 m; integument black, clypeus, apical protarsomeres and apical mesotarsomeres yellow; vestiture dark brownish-black in color, setae at apex of abdomen entirely black; setal fringe of protibiae and mesotibiae bright yellow; wings hyaline with feeble greenish-blue iridescence.

Notes: The specimen of *X. torrida gramineipennis* that I examined did not differ significantly from the other material of this species that I examined for this study.

Additional Material Examined: CAMEROON: Medzek, Akonolinga, 18.XI.2003, J. Mbida, ex flowers (14 females and 1 male). CÔTE D'IVOIRE: Abidjan, 18-26.VIII.1952, L. Sheljuzhko (1 female). DEMOCRATIC REPUBLIC OF THE CONGO [Belgian Congo]: Kimnenza, A. Schultze, 22-26.IX.1910 (1 female); Luebo, D. W. Snyder (2 females). NIGERIA: Ibadan, 7-10.III.1929, C. B. Phillip (1 female), same locality but 29.VI.1964, J. C. Ene (1 female); Olokemeji, Ibadan (11 females). Eardley (1987) also records this species from: Angola, Botswana, Equitorial Guinea, Ethiopia, Gabon, Ghana, Liberia, Rwanda, Sierra Leone, Tanzania, and Uganda.

Giant Carpenter Bees

Chapter 4: Asian Species

The Asian species included by Michener (2007) in the subgenus *Mesotrichia* belong to three morphologically distinct groups. In the earlier classification of the genus *Xylocopa* proposed by Hurd and Moure (1963), each of these groups was recognized as a distinct subgenus. Two of these groups, which correspond to the subgenera *Hoplitocopa* and *Hoploxylocopa* of Hurd and Moure (1963), contain only a single species. The other group, corresponding to the subgenus *Platynopoda* Westwood as defined by Hurd and Moure (1963), includes three species. As there are unique morphological features which can be used to define each of these three groups, I follow Hurd and Moure (1963) in recognizing these three subgenera among the Asian species of the "Mesotrichia Group."

The following key will separate the known Asian species of all three subgenera included in the "Mesotrichia Group."

Key to Asian Species of the "Mesotrichia Group"

1) Lateral margins and apex of metasoma with elongate reddish-orange setae... *X. assimilis*
- Lateral margins and apex of metasoma with elongate black setae... 2
2) Wing iridescence uniformly bronzy-coppery or greenish-bronzy, iridescence brilliant in females but less bright and distinct in males... *X. acutipennis*
- Wing iridescence other colors, wings strongly iridescent in both sexes... 3
3) Wing iridescence green and/or blue basally and violet or reddish-violet apically... *X. latipes*
- Wing iridescence green basally, becoming pinkish or pinkish-green apically... *X. perforator*
- Wing iridescence blue basally and green or greenish-gold apically .... *X. tenuiscapa*

Subgenus *Hoplitocopa* Hurd and Moure (1963:257)

Diagnostic Features: Individuals of both sexes with elongate reddish-orange setae along lateral margins of metasoma; mandibles and labrum greatly elongated in male; mandibles slightly elongated in female.
Type species: *Xylocopa assimilis* Ritsema, by original monotypy.

Xylocopa (Hoplitocopa) assimilis
Ritsema (1880:221)

Figure 10: a: female, dorsal view; **b:** female, head, frontal view, showing elongated mandibles; **c:** male, dorsal view; **d:** male, head, frontal view, showing elongated mandibles and labrum; e: female, metasoma, showing lateral and apical setae; **e:** male, metasoma, showing lateral and apical setae; **f:** lateral projection of metatibia; **g:** distribution map (green stars).

Xylocopa assimilis Ritsema (1880:221), type locality "Sumbawa," species concept based on the two paratypes of *X. gastrica* Maa in USNM which had subsequently been identified by M. A. Lieftinck and P. D. Hurd as specimens of *X. assimilis*.

Xylocopa gastrica Maa (1939:95), synonymy by Lieftinck (1955:9, 29), type locality "N. W. Soemba: Laora, 100 m," type material examined: PARATYPE, male, "Bondo Kodi, W. Sumba," and PARATYPE, female, "Lokojengo, C. Sumba" (USNM).

Brief Diagnosis: Female: body large, black, overall body length 26 mm, metasomal width 12 mm, mesosomal disc glabrous and strongly shining, disc of metasomal tergites glabrous, mesosoma laterally with dense black pubescence, metasoma with a fringe of reddish-orange setae along the lateral and apical margins, wings subhyaline with feeble greenish-bronzy iridescence, mandibles slightly elongated relative to other species of "Mesotrichia Group." **Male:** similar to female in overall appearance, overall body length 25 mm, metasomal width 12 mm, labrum elongate, mandibles elongate and slender, mandibles and labrum both much longer than those of other species in the "Mesotrichia Group."

Subgenus *Hoploxylocopa* Hurd and Moure (1963:260)

Diagnostic Features: Female with brilliant bronzy-coppery iridescence on the wings; male wings hyaline but with bronzy-coppery reflections. Male with an elongate spine on the inner margin of the hind trochanter (second segment of the leg) and a well developed gradulus or impressed area on the third metasomal tergite.
Type species: *Xylocopa acutipennis* Smith, by original monotypy.

Xylocopa (Hoploxylocopa) acutipennis Smith (1854:355)

Figure 11: a: female, with bronzy-coppery wing iridescence, dorsal view; **b:** female, with greenish-bronzy wing iridescence, dorsal view; **c:** male, dorsal view; **d:** distribution map, by country.
Xylocopa acutipennis Smith (1854:355), type locality "Silhet," type material: HOLOTYPE, male, "Silhet" (BMNH).
Xylocopa splendidipennis Ritsema (1876:183), synonymy by Hurd and Moure (1963:316), type locality "Sumatra."
Mesotrichia kerri Cockerell (1929:303), new synonymy, type locality "Siam, Kao Pawta, Changotong, Ranawg," type material examined: HOLOTYPE, female, "Fls Melastoma, Kao Pawta, Changotong, Ranawg, Siam. Jan 21., (Kerr.)" (USNM).
Brief Diagnosis: Female: moderately large-bodied, overall body length 22-29 mm, metasomal width 11-13 mm, integument black, vestiture black, mesosoma with dense black pubescence laterally, dorsal surface of metasomal tergites with short black pubescence and longer black setal tufts laterally, wings subhyaline with brilliant, unformly colored orange-red or greenish-bronzy iridescence. **Male:** smaller-bodied, overall body length 19 mm, metasomal width 10 mm, integument black with black vestiture, anterior tibiae and tarsi with

lateral fringe of white or grey setae, third metasomal tergite with distinct gradulus, wings hyaline with feeble bronzy or greenish-bronzy iridescence.

Notes: The holotype of *Mesotrichia kerri* Cockerell represents the female of the species which Smith had previously described as *Xylocopa acutipennis* based on a male specimen. Smith's name has clear priority and *Mesotrichia kerri* must therefore be placed in synonymy with *Xylocopa acutipennis*.

Additional Material Examined: INDIA: Khasia Hills (1 male). INDONESIA: Sumatra, Fort de Kock, VIII.1918 (1 male). MALAYSIA: Perak: Taiping (1 female). MYANMAR: Carin Cheba, 900-1100 m, L. Fea, V-XII.1889 (1 female). NEPAL: Katmandu Valley, Nagarjun Forest (1 female). The website DiscoverLife.org provides additional localities for this species in Bangladesh, China, India, Laos, Myanmar, and Thailand.

Subgenus *Platynopoda* Westwood (1840:271)

Diagnostic Features: Body size massive in both sexes; wings brownish-black to black with strong metallic iridescence; females with a strong transverse carina across the scutellum; males with first segment of the first pair of tarsi flattened and/or greatly expanded.

Type Species: *Apis latipes* Drury, subsequent designation by Ashmead (1899:71).

Notes: Maa (1940) published a revision of the species of the subgenus *Platynopoda* but unfortunately his key to species includes a number of significant errors, particularly in regards to the dimensions of the first segment of the first pair of tarsi in the males, which make it extremely difficult to accurately identify species using the key. Maa's key was republished with some modifications by Hurd and Moure (1963), but unfortunately these authors did not correct the errors regarding the male tarsal segments.

Xylocopa (Platynopoda) latipes
(Drury 1773:87)

Figure 2: females visiting flowers of *Thunbergia grandiflora* var. *alba* Leonard (Acanthaceae), photographed October 27, 2017, by the author on Corregidor Island, Philippines.

Figure 12: a: female, dorsal view; **b:** male, dorsal view; **c:** distribution map, by country except for Indonesia and Philippines which are mapped by island or island group.

Apis latipes Drury (1773:87), type locality "Island of Johanna," type material not traced and according to Maa (1940:572) probably lost. Species concept based on illustration of male type specimen provided by Drury (1773:pl. 48, f. 2).

Apis gigas DeGeer (1773:576), type locality "Inde," type material not traced and according to Maa (1940:572) probably lost.

Mesotrichia (Platynopoda) latipes basiloptera Cockerell (1917:347-349), type locality "Puerto Princesa, Palawan," type material examined: HOLOTYPE, female, "P. Princesa, Palawan, Baker" (USNM).

Brief Diagnosis: Female: body massive, overall body length 27-38 mm, metasomal width 11-14 mm, integument black, vestiture black, disc of mesonotum glabrous and strongly shining, disc of metasoma lacking dorsal setae and finely, densely punctate, lateral portions of mesosoma with dense black pubescence, lateral margins of metasomal segments with elongate black setal tufts, wings dark brownish-black with strong iridescence: green or blue basally and violet or reddish-violet on apical field. **Male:** body massive, overall body length 23-33 mm, metasomal width 12-13 mm, compound eyes greatly enlarged and covering more than half of head capsule, integument black, first tarsal segment of the front legs whitish-yellow, flattened, and s-shaped, tarsi of the front legs with elongate yellowish-white setae laterally, apical tarsi with elongate black setae, wings colored as in female.

Additional Material Examined: INDIA: Assam, W. F. Badgley, 1906-185, from Brit. Mus. (1 male). INDONESIA: Borneo: North Borneo, E. Coast Residency, Kinabafangan Dist., nr. Mouth Little Kretam River, in burrows in approx. 25" log, 7.V.1950 (1 female, 1 male); Sandakan, Baker (1 female, 1 male); West Kalimantan: Mt. Bawang (1 female). Java: Buitenzorg, III.1909, Bryant and Palmer (4 females, 1 male), same data except IV.1909 (1 female, 1 male), Apr.-Dec. 1896, D. G. Fairchild (5 females); Depok, 8.I.1909, Bryant and Palmer (1 female), same data except X.1909 (1 female); West Java, Djakarta, 18.II.1957, C. L. Klein (1 female); Jakarta, Cilandak, 24.XI.1986, B. A. Annis (1 male); Mt. Salak, Java, 3.VII.1909, altitude 2500 feet, Bryant and Palmer (2 females); Pelabocan, Ratoe, Java, 12.X.1909, Bryant and Palmer (1 female); Sockaboemi, Java, 15.I.1909, Bryant and Palmer (2 females, 1 male). Sumatra: West Sumatra, Bandar Buar Padang, H. Schode (1 male); Fort de Kock, 920 m, 1925, E. Jacobson (1 female, 1 male); Jambi, Sungai Bengkal, 9.XII.1977, D. Pletsch (1 female); Kephalang, 1960 feet, Nov-Dec 1923, H. C. Kellers (6 females, 2 males); Kampong, Silau, Maradj, Asaham, 1918, H. H. Bartlett (1 female, 1 male (dissected by Hurd)). Sumbawa: Musa Tenggara Barat Prov., Ambalawai River, 21.X.1985, J. D. Weintraub (1 female). MALAYSIA: Island of Penang, Baker (3 females, 2 males); Perak: 10 m of Ipoh in forest, R. Traub, elev. 150 ft. (1 male); Perak: Tapah Hill (6 females, 7 males); Perak, 1987 (1 female, 1 male). Sabah: Apin Apin, 2.IX.1983, G. F. Hevel and W. M. Steiner (1 male); 10 km S Matunggong, 19.IX.1983, G. F. and J. F. Hevel and W. E. Steiner (1 male); Weston, 1.VIII.1983, W. E. Steiner, G. F. Hevel (1 female.). Sarawak: Semengoh Forest Reserve, 15 miles S. Kuching, 17.IX.1966 (2 females). Selangor: 16 miles N. of Kuala Lumpur, III.1949 (1 female); Kuala Lumpur, 1950 (1 male). MYANMAR: Tenasserim, Thagata, Fea, IV.1887 (1 male); "Burmah" (1 female). NEW GUINEA: Lae, 5.VIII.1972, D. H. Messersmith (1 male). PHILIPPINES: Brooke's Py, 18.XII.1965, D. Davis (1 female); Culasi Panay June 1913 McGregor (1 female, 1 male); Cuernos Mts., Negros, Baker (3 females); Irawan, 14 km W. P. Princesa, 11-13.XII.1965, D. Davis (1 male); 28 km W P. Princesa, Chromite Mine, 1-6.XII.1965, D. Davis (1 male); Negros, 10.X.1925 (1 female); Puerto Princesa Palawan, 19.VIII.1925, R. C. McGregor (1 female, 1 male), same data except 23.VIII.1925 (1 male), same data except VIII.1925 (1 female); P. Princesa, Palawan, Baker (5 females). SINGAPORE: East Coast Forest, 2 km E Singapore, M. Foo (1 male); without further data (1 female, 1 male). THAILAND:

Bangkok, X.1924 (1 female); Bangkok, H. Smith collector, 22.VI.1928 (3 females), same data except 29.VI.1926 (4 females), no date (2 female); Kaosabab, 500 m, 24.XI.1933, H. M. Smith (1 female); Ko Chang Island (1 female); Ko Pha Ngan, "Gulf of Siam," 24.VII.1931, H. Smith (1 male); Mae Tha, Lamphun (1 female, 6 males); Nakon, Sritamarat, Siam, H. Smith 16.VII.1928 (1 male); Singora, Siafra, Siam, VI.1929, H. M. Smith (2 females); Prov. Nakhon, Ratchasima, Sakaerat Expt. Sta. 2-4.III.1971, C. Schneider (2 females) 14 30 N, 101 55 E, 300-600 m; Trong, "Lower Siam," Dr. W. L. Abbott (20 females, 6 males). VIETNAM: Central Highlands, Pleikee area, VI.1968, V. K. Milbank (1 male); "Cochin China," C. F. Baker Collection (1 female). The website DiscoverLife.org also provides additional localities for this species in Bangladesh, Cambodia, China, Laos, and Sri Lanka as well as a possible locality on Ternate which is based on a photographic record only. The New Guinea record above is potentially doubtful, as this locality occurs well outside the known distribution of the species.

Xylocopa (Platynopoda) perforator
Smith (1861:61)

Figure 13: a: female, dorsal view; **b:** male, dorsal view; **c:** distribution map.

Xylocopa perforator Smith (1861:61), type locality "Ternate," type material: SYNTYPES, male and female, "Ternate" (BMNH).

Brief Diagnosis: Female: body massive, overall body length 24-32 mm, metasomal width 13-15 mm, similar to *X. latipes* in size and general appearance but with distinctly different wing iridescence, which is greenish basally and which becomes either pink or pinkish-green on the apical field. **Male:** body massive, overall body length 24-28 mm, metasomal width 11-13 mm, similar to *X. latipes* in general appearance but with the first tarsal segment of the front pair of legs not as strongly expanded, more rectangular in shape, and the wing iridescence distinctly different, greenish basally and becoming pink or pinkish-green on the apical field.

Additional Material Examined: INDONESIA: Java: Buitenzorg, IV-XII.1896 (2 females), III.1904 (2 females, 1 male), 1906-1907 (3 females), IV.1909 (2 females), no date (1 female); Depok, VII.1909 (4 females); Sindanglaya, 27.XI.1920 (2 females); "East Java" (5 females, 6 males). The website DiscoverLife.org also provides additional localities for this species in Indonesia (Lombok, Flores, Sumatra, and Timor), as well as Malaysia (North Borneo).

Xylocopa (Platynopoda) tenuiscapa
Westwood (1840:271)

Figure 14: a: female, dorsal view; **b:** male, dorsal view; **c:** distribution map, by country.

Xylocopa (Platynopoda) tenuiscapa Westwood (1840:271), type locality "India," type material: HOLOTYPE, male, Oxford University Museum of Natural History.

Xylocopa viridipennis Lepeletier (1841:205), synonymy by Maa (1940:569), type locality "Inde," type material: HOLOTYPE, female, Turin Museum.

Xylocopa latreillei Lepeletier (1841:206), synonymy by Maa (1940:570), type locality "Bengale," type material: SYNTYPE, male, Oxford University Museum of Natural History.

Xylocopa lativentris Blanchard (1844:30), synonymy by Maa (1940:570), type locality not stated in original description but likely Kashmir according to Maa (1940:570), type material: SYNTYPES, females, Museum National d'Histoire Naturelle, Paris.

Xylocopa albofasciata Sichel (1867:154), synonymy by Maa (1940:570), type locality "Taprobane, Ceylon," type material: HOLOTYPE, female, Vienna Museum.

Mesotrichia latipes var. *magnifica* Cockerell (1929:302), new synonymy, type locality "Doi Sutep, Siam," type material examined: HOLOTYPE, female, "Doi Sutep, Mch 23-28, Siam. (McKean)" (USNM).

Brief Diagnosis: Female: body massive, overall body length 24-34 mm, metasomal width 13-16 mm, similar to *X. latipes* but differing most conspicuously in the coloration of the wing iridescence, which is brilliant blue or blue-violet at base and green or golden-green on the apical field. **Male:** body massive, overall body length 24-29 mm, metasomal width 12-14 mm, similar to *X. latipes* in general appearance but differing notably in its wing iridescence, which is brilliant blue or blue-violet at base and green or golden-green on the apical field.

Notes: The holotype female of *Mesotrichia latipes* variety *magnifica* Cockerell is housed in USNM and images of this specimen are currently available online at the USNM website. I also had the opportunity to study this specimen in person at USNM in 2016 and 2017. Maa (1938:320, 324 and 1940:567, 572) considered *X. magnifica* to be a valid species based on Cockerell's description, but it is doubtful that he ever actually examined the type specimen. My own personal examination of this specimen has convinced me that it is in every way a normal female specimen of *X. tenuiscapa* except for the slightly shorter third antennal segment. The specimen does not otherwise differ from additional specimens of *X. tenuiscapa* which were collected in Thailand, including material collected at or near the type locality for *X. magnifica* (Doi Sutep, which is immediately west of Chiang Mai).

Additional Material Examined: CHINA: Yun Hsien, II.1942, W. L. Jellison (1 male), same data except III.1942 (1 female), same data except IV.1942 (1 male). INDIA: Assam: Chabua, VII.1943, W. L. Jellison (2 males); Doom Dooma, 23.V.1945, D. E. Hardy (1 female), same data except VI.1943 (3 females, 1 male); "Assam" (1 female). Goa: Mormugao (2 females). Jharkand: Ranchi, IV.1957 (1 female). Karnataka: Mudigere (19 km W), 6.IV.1980, Mathis and Freidberg (1 female); South Coorg, Ammathi, XI.1950 (1 female), 17.V.1951 (1 female). Kerala: Kerala Survey, 12.5 km W Pechiparai, M. S. Mani and party, 15-27.VIII.1974 (1 female). Maharashtra: Bombay, Lonavla, 2000 ft., 10.V.1958, F. L. Wain (1 male). Odisha: Badrama, 1100', 20-26.II.1975, M. L. Ripley (1 male); Simlipai Hills, Meghasini Hill, II.1975, 3700', M. L. Ripley (1 male). Pudicherry: Karikal, VIII.1958, P. S. Nathan (1 female, 1 male), same data except XII.1958 (3 males), same data except 24.I.1959 (1 male), same data except II.1959 (2 males), same data except VII.1965, P. S. Nathan (1 male), same data except II.1966, P. S. Nathan (2 females), XI.1989,

Giant Carpenter Bees

T. R. S. Nathan (2 females). Tamil Nadu: Anamalai Hills, Kadenparai, 3500', V.1963, P. S. Nathan (3 females). Coimbatore, III.1954, P. S. Nathan (1 male), same data except X.1958 (2 males), same data except 1400 feet, VII.1963, P. S. Nathan (72 females, 3 males), same data except XI.1962 (1 female), same data except VI.1963 (79 females), same data except VIII.1963 (2 females), same data except VIII.1964 (100 females), same data except VII.1965 (1 male), same data except IX.1965 (4 males), same data except X.1965 (4 males), same data except XI.1965 (1 male), same data except II.1966 (1 female). Telangana: Hyderabad, 10.X.1968 (1 female). West Bengal: Calcutta, 28.XI.1973, M. L. Ripley (1 male). "India" (2 females). NEPAL: Jhapa District, Tea Garden, 400 m, 3.IV.1967, D. Nicolson (1 male); Kailali Dist., Malaketi, 14 m. NW Dhangarhi, 200 m., 30.XI.1966 (1 female). SRI LANKA: Anuradhapura District: Irrigation Bungalow, Padaviya, 180 ft., 27.II-9.III.1970, Davis and Rowe (1 female); Wilpattu Nat'l. Park, Panikka Wila Bungalow, 1.XI.1977 (1 male). Badulla District: Ella Resthouse, 3400 ft., blacklight, 17-20.XI.1974 (2 females); Madulsima, VIII.1918, T. B. Fletcher (1 male). Colombo District, Kalatuwawa, 12-15.VIII.1975 (1 female, 1 male); Labugama Reservoir, 400 ft. elevation, 2-3.X.1976 (4 females); Labugama Reservoir, Jungle, 13-14.X.1973 (1 female); Malawana, sea level, 22.VIII.1973, G. Ekis, collected at black light (2 females); Ratmalana, Zoo Farm 19.I.1975 (1 male). Galle District: Kaneliya, 500 feet, 21-22.IV.1973, Baumann and Cross, at black light (1 female); Kanneliya Forest, Black Light, 16.V.1974, Gans and Prasanna (2 females). Hambantota District: Palutapana, 10-12.VIII.1972 (2 females, 1 male); WLNPS Bungalow 0-15 m, 27-29.IX.1977 (1 female). Kandy District: Kandy, 21.II.1971, Piyadasa and Somapala (1 female, 1 male); Katugastota, 2.X.1973 (1 male); Madugoga, ca. 2600 ft., 1-IV.1973, at black light, Baumann and Cross (1 male); 5 miles N Mahiyangana, 30.III-9.IV.1971, P. and P. Spangler (1 female); Peradeniya, 12.X.1903, W. F. H. Rosenberg (1 male), 1-15.II.1971, Piyadasa and Somapala (2 females, 1 male). Kurunegala District: Kurunegala, Athugala (Elephant Rock), 24-25.I.1975 (1 female, 1 male). Mannar District: Olaithoduvai, 10 mi NW of Mannar, 0-50 feet, 4-5.XI.1976 (2 males). Matale District: Sigiriya, 800 ft., 25.II.1970, Davis and Rowe (1 male), 25.XI.1974 (1 female). Polonnaruwa District: Sigiriya, 3.III.1972 (1 male). Puttalam District: Wilpattu National Park, Kali Villu, 12-14.VI.1975 (1 male). Ratnapura District: Ratnapura Rest House, 24.IX.1977 (1 male); Singharaja Forest, 5.VIII.1973, 600 feet, G. Ekis, collected at black light (1 female); Uggalkaltota, 350 ft., Irrigation Bungalow, 31.I-8.II.1970, Davis and Rowe (1 female). THAILAND: Bangkok, 10.IX.1929, H. Smith (1 male), 10.V.1934, S. Tongyai (1 male), 18.V.1979, P. Schaefer (1 male); Chiangmai, 18.X.1952 (1 female); Chiang Mai (8 females, 2 males); Doe Suthep, Chiang Mai, 29.V.1988 (1 male); Doi Angka, 5.XII.1928, H. M. Smith (1 male), same data except 22.VI.1928 (1 female); Fang, Chiang Mai (1 female, 1 male); Mae Tha, Lamphun (4 females, 16 males). VIETNAM: Saigon, "Cochin China" (1 female). The DiscoverLife.org website also includes information on specimens collected in Cambodia and Laos. The distribution maps of Maa (1940) indicate that this species also occurs in Bangladesh and Myanmar.

Doubtful Species of Subgenus *Platynopoda*

As noted in the Introduction, other bees and insects (including non-stinging insects such as flies) resemble certain species of the "Mesotrichia Group" in their general body proportions, coloration, and wing iridescence. In particular, several species of the subgenus *Biluna* Maa of the genus *Xylocopa* closely resemble the three large, common Asian species of the subgenus *Platynopoda* and are often confused with these species in museum collections and in the global specimen trade. I have illustrated three of the most common and frequently confused species of subgenus *Biluna* in Figure 15.

The identity of one species now included in subgenus *Platynopoda*, *Xylocopa marginella* Lepeletier (1841:205, described from "Java"), has long been a subject of debate (Smith 1874; Maa 1940; Hurd and Moure 1963). Smith (1874) thought that *X. marginella* was probably a synonym of *X. tenuiscapa*, while Maa (1940) and Hurd and Moure (1963) treated *X. marginella* as a separate species within the subgenus *Platynopoda,* based solely on Lepeletier's mention of reddish-brown pubescence at the base of the hind tarsi. However, such pubescence occurs occasionally in specimens of all of the species of subgenus *Platynopoda*, indicating that this character alone is not sufficient for recognizing *X. marginella* as a separate species. There are no specimens identified as *X. marginella* in USNM, and the treatments by Smith (1874), Maa (1940), and Hurd and Moure (1963) were all apparently based on the original description alone. The original, brief description of *X. marginella* by Lepeletier (1841:205) does not contain sufficient information to accurately determine which of several similar sympatric species is actually indicated. One possible interpretation is suggested by the fact that the original description mentions an apical band of violet iridescence on the wings, a characteristic which is found in certain species of subgenus *Biluna*, particularly *X. (Biluna) iridipennis* Lepeletier (see female specimen illustrated in Figure 15a) which is common on Java. Based on this feature in the original description, it is possible that the name *X. marginella* may refer to the species now known under the name *X. (Biluna) iridipennis*. Another possibility is that the name *X. marginella* is a senior synonym of *X. perforator* Smith, a species which also occurs on Java and whose features match many of those contained in the very brief decription of *X. marginella* by Lepeletier (1841). Given the significant uncertainty regarding the identity of *X. marginella,* any formal proposal of synonymy must await rediscovery and examination of the type specimen(s) of *X. marginella*. At this time I believe it is best to consider the name *X. marginella* a *species incertae sedis* until the type material of this species is rediscovered.

Wu (1982) described a new species of subgenus *Platynopoda* from southern China, *Xylocopa (Platynopoda) yunnanensis*. Based on the original description and information provided in the publication by Wu (1982), I believe that this species should be transferred to the subgenus *Biluna*, where it is possibly a synonym of *X. (B.) nasalis* Westwood, a species with blue iridescent wings which is known from southern China (Hurd and Moure 1963).

Figure Captions

Figure 1: *Xylocopa flavorufa:* females visiting flowers of *Karomia speciosa* (Hutchinson & Corbishley) R. Fernandes (Lamiaceae), photographed November 26, 2017, by the author at Skukuza, Mpumalanga, Republic of South Africa, in the Kruger National Park.

Figure 2: *Xylocopa latipes:* females visiting flowers of *Thunbergia grandiflora* var. *alba* Leonard (Acanthaceae), photographed October 27, 2017, by the author on Corregidor Island, Philippines.

Figure 3: *Xylocopa chapini:* **a:** female PARATYPE, dorsal view; **b:** male PARATYPE, dorsal view; **c:** female, front of head; **d:** female, metasoma, showing apical setae; **e:** distribution map, by country.

Figure 4: *Xylocopa combusta:* **a:** female, dorsal view; **b:** male, dorsal view; **c:** female, metasoma, showing apical setae; **d:** distribution map, by country.

Figure 5: *Xylocopa flavorufa:* **a:** female, compared with type by P. D. Hurd, with extensive red pubescence on mesosoma, dorsal view; **b:** female, with reduced red pubescence on mesosoma, dorsal view; **c:** male, dorsal view; **d:** female, front of head; **e:** female, metasoma, showing apical setae; **f:** distribution map, by country.

Figure 6: *Xylocopa ignescens:* **a:** female, compared with type, dorsal view; **b:** male, compared with type, dorsal view; **c:** female, metasoma, showing apical setae; **d:** distribution map, by country.

Figure 7: *Xylocopa mixta:* **a:** female, with reduced red pubescence on mesosoma, dorsal view; **b:** female, with extensive red pubescence on mesosoma, dorsal view; **c:** male, dorsal view; **d:** female, front of head; **e:** female, metasoma, showing apical setae; **f:** distribution map, by country.

Figure 8: *Xylocopa subcombusta:* **a:** female PARATYPE, dorsal view; **b:** male, PARATYPE, dorsal view; **c:** female, metasoma, showing apical setae; **d:** distribution map, by country.

Figure 9: *Xylocopa torrida:* **a:** female, dorsal view; **b:** male, dorsal view; **c:** female, metasoma, showing apical setae; **d:** distribution map, by country.

Figure 10: *Xylocopa assimilis:* **a:** female, dorsal view; **b:** female, head, frontal view, showing elongated mandibles; **c:** male, dorsal view; **d:** male, head, frontal view, showing elongated mandibles and labrum; **e:** female, metasoma, showing lateral and apical setae; **e:** male, metasoma, showing lateral and apical setae; **f:** lateral projection of metatibia; **g:** distribution map (green stars).

Giant Carpenter Bees

Figure 11: *Xylocopa acutipennis:* **a:** female, with bronzy-coppery wing iridescence, dorsal view; **b:** female, with greenish-bronzy wing iridescence, dorsal view; **c:** male, dorsal view; **d:** distribution map, by country.

Figure 12: *Xylocopa latipes:* **a:** female, dorsal view; **b:** male, dorsal view; **c:** distribution map, by country except for Indonesia and Philippines which are mapped by island or island group.

Figure 13: *Xylocopa perforator:* **a:** female, dorsal view; **b:** male, dorsal view; **c:** distribution map.

Figure 14: *Xylocopa tenuiscapa:* **a:** female, dorsal view; **b:** male, dorsal view; **c:** distribution map, by country.

Figure 15: Species of *Xylocopa* subgenus *Biluna* from Southeast Asia and Indonesia: a: *Xylocopa (Biluna) iridipennis* Lepeletier, female, dorsal view; **b:** *Xylocopa (Biluna) auripennis* Lepeletier, female, dorsal view; **c:** *Xylocopa (Biluna) nasalis* Westwood, female, dorsal view. These species are frequently confused with species of the "Mestotrichia Group" in museum collections and in the global insect specimen trade.

Figure 1: *Xylocopa flavorufa*

Giant Carpenter Bees

Figure 2: *Xylocopa latipes*

Figure 3: *Xylocopa chapini*

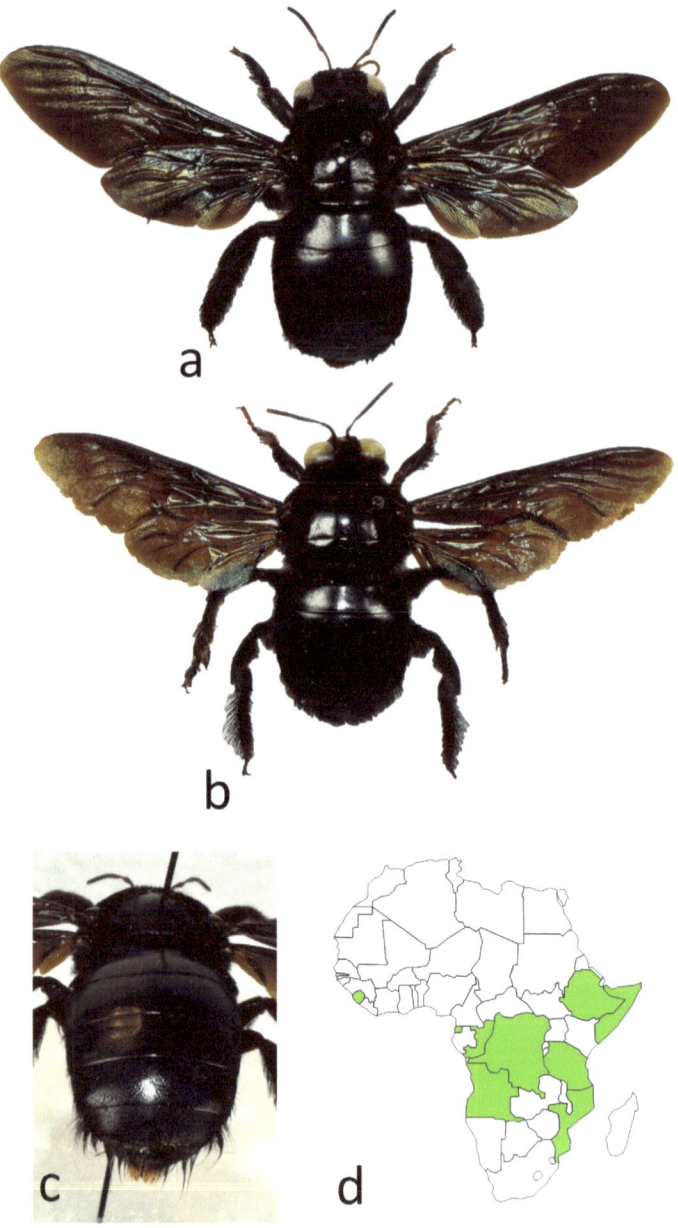

Figure 4: *Xylocopa combusta*

Figure 5: *Xylocopa flavorufa*

Giant Carpenter Bees

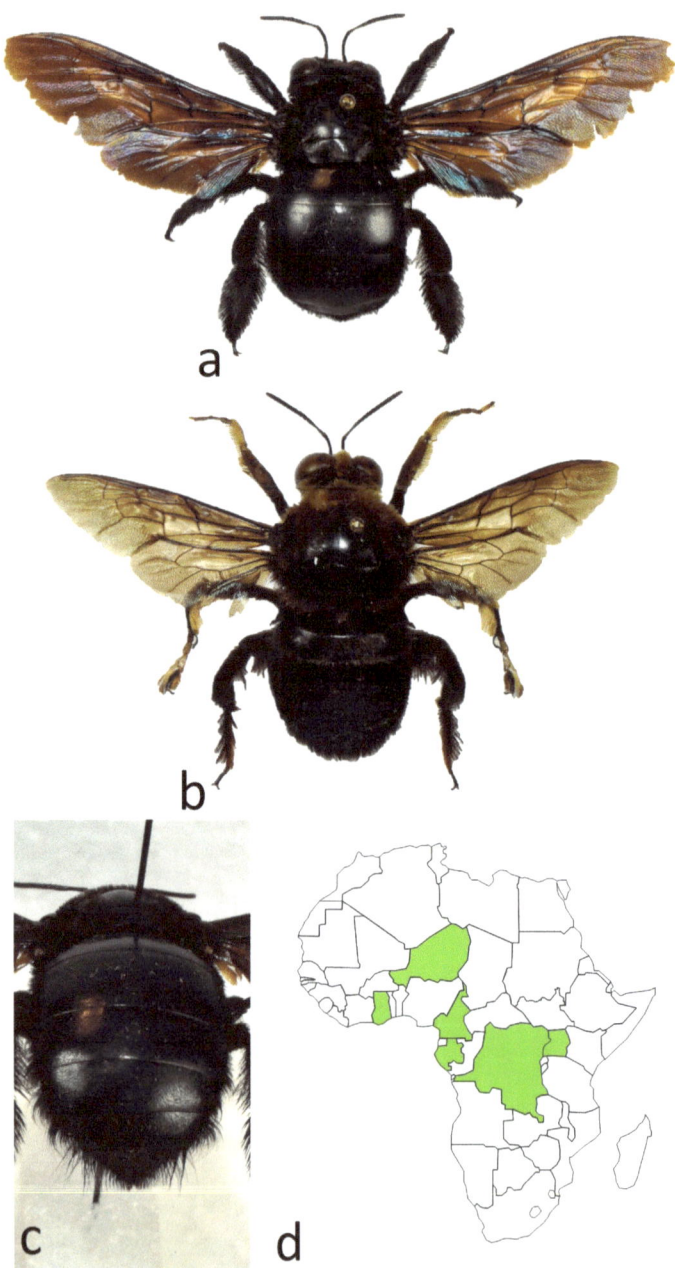

Figure 6: *Xylocopa ignescens*

Figure 7: *Xylocopa mixta*

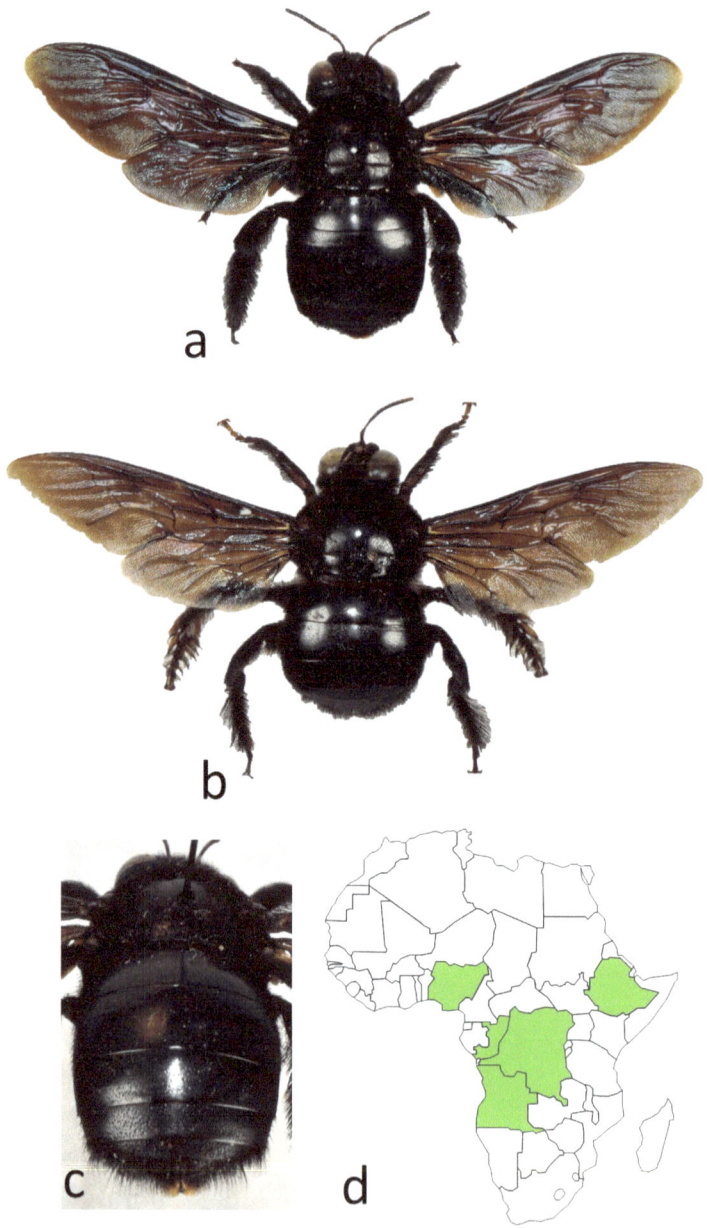

Figure 8: *Xylocopa subcombusta*

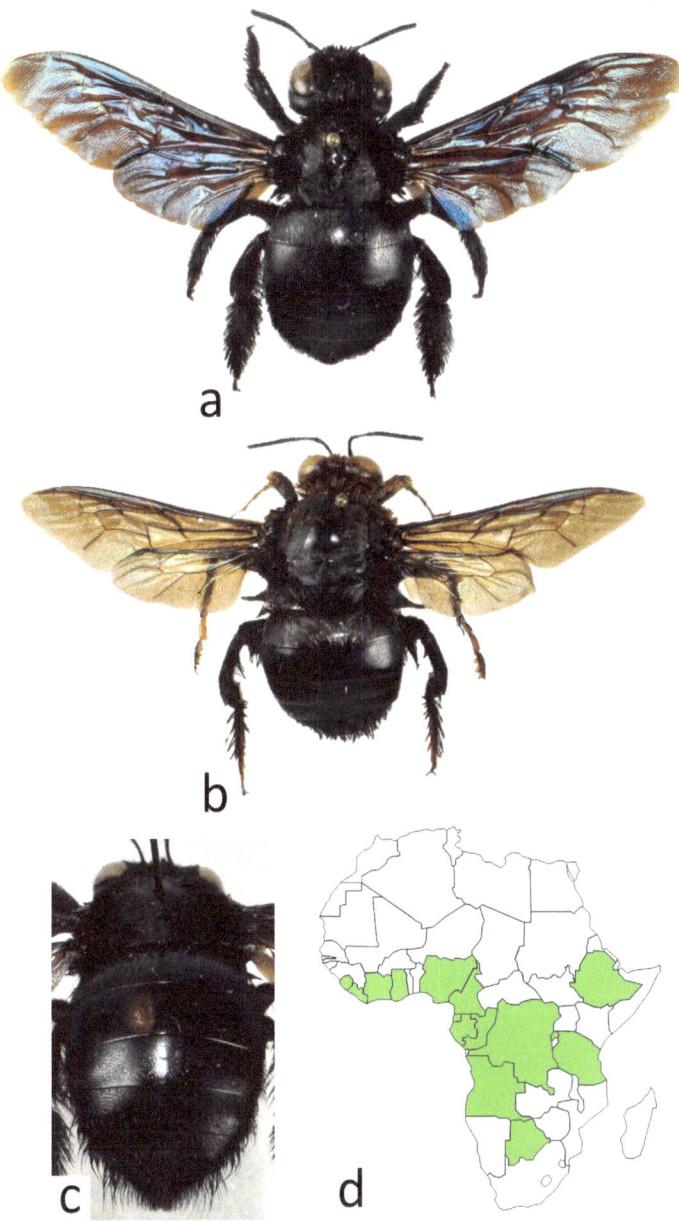

Figure 9: *Xylocopa torrida*

Giant Carpenter Bees

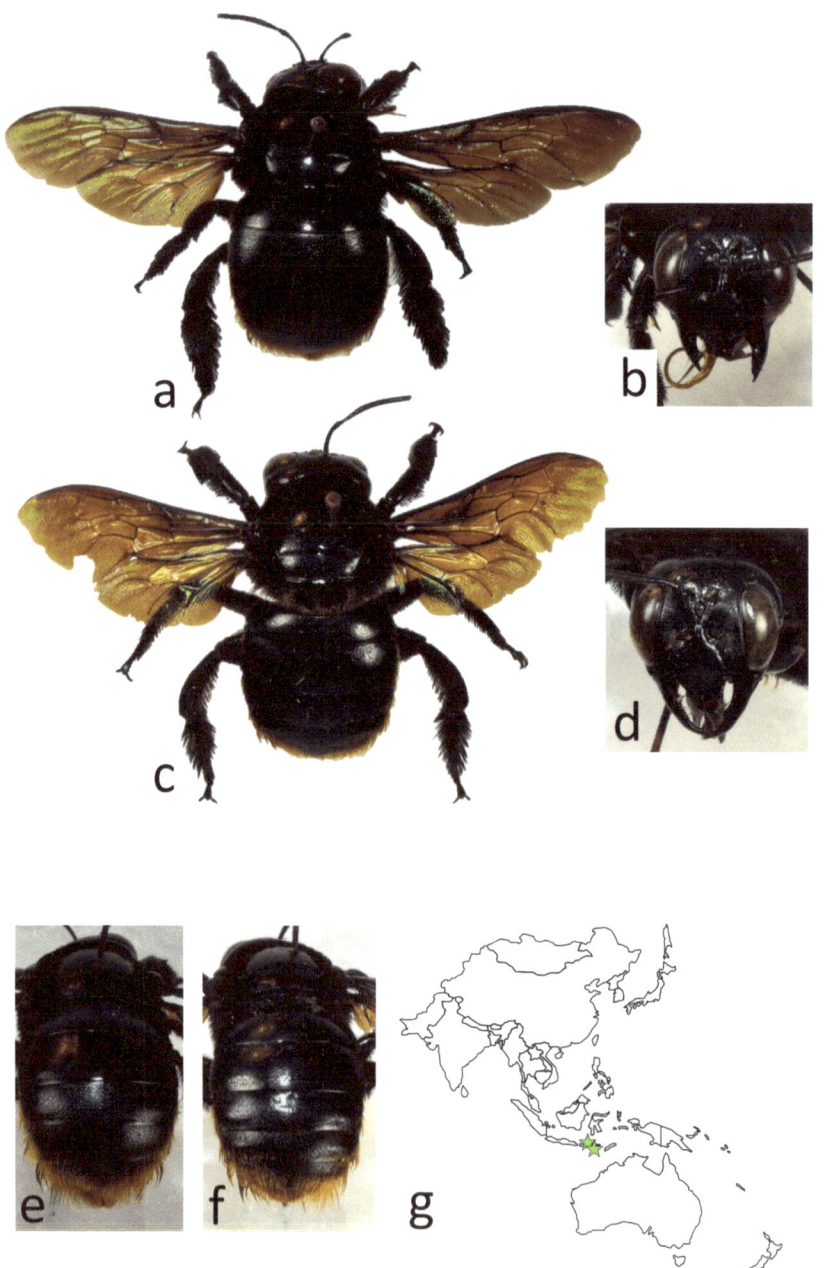

Figure 10: *Xylocopa assimilis*

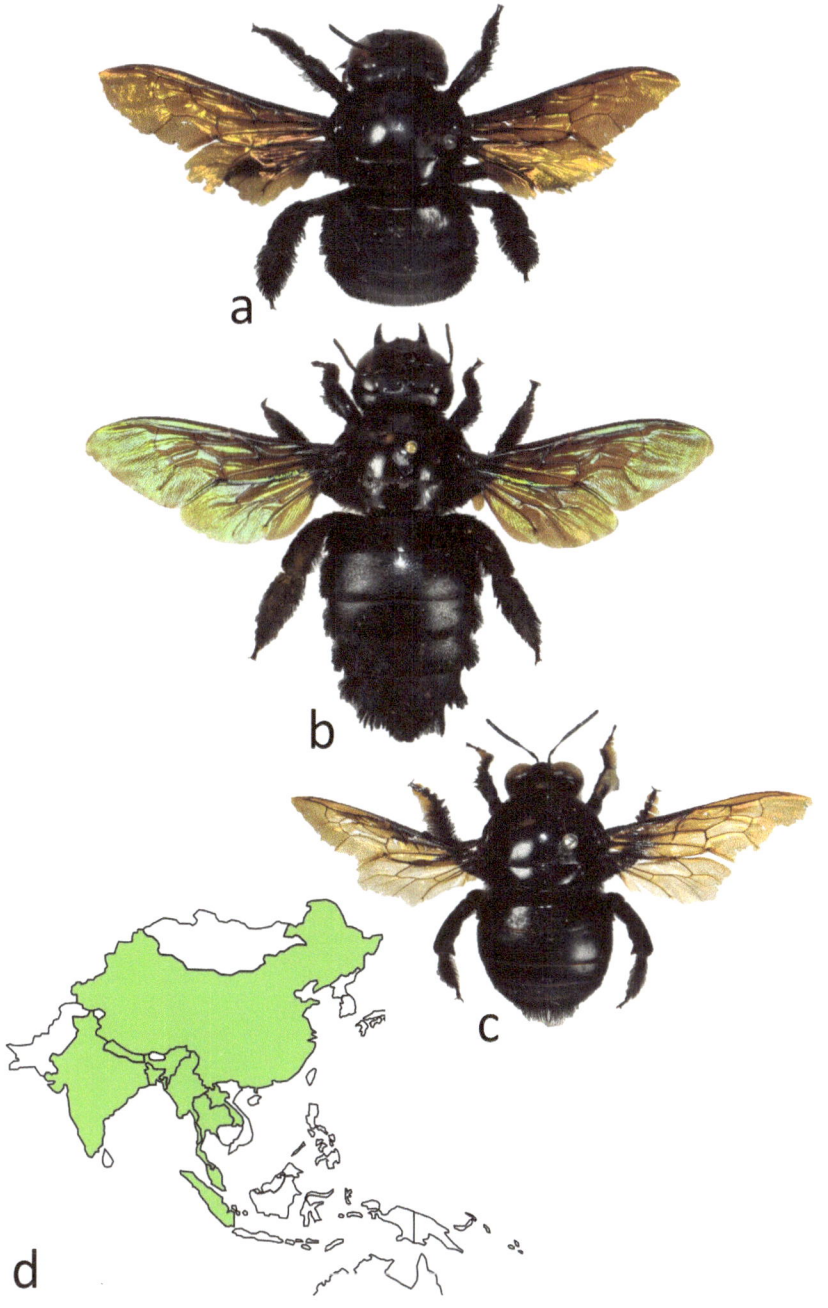

Figure 11: *Xylocopa acutipennis*

Giant Carpenter Bees

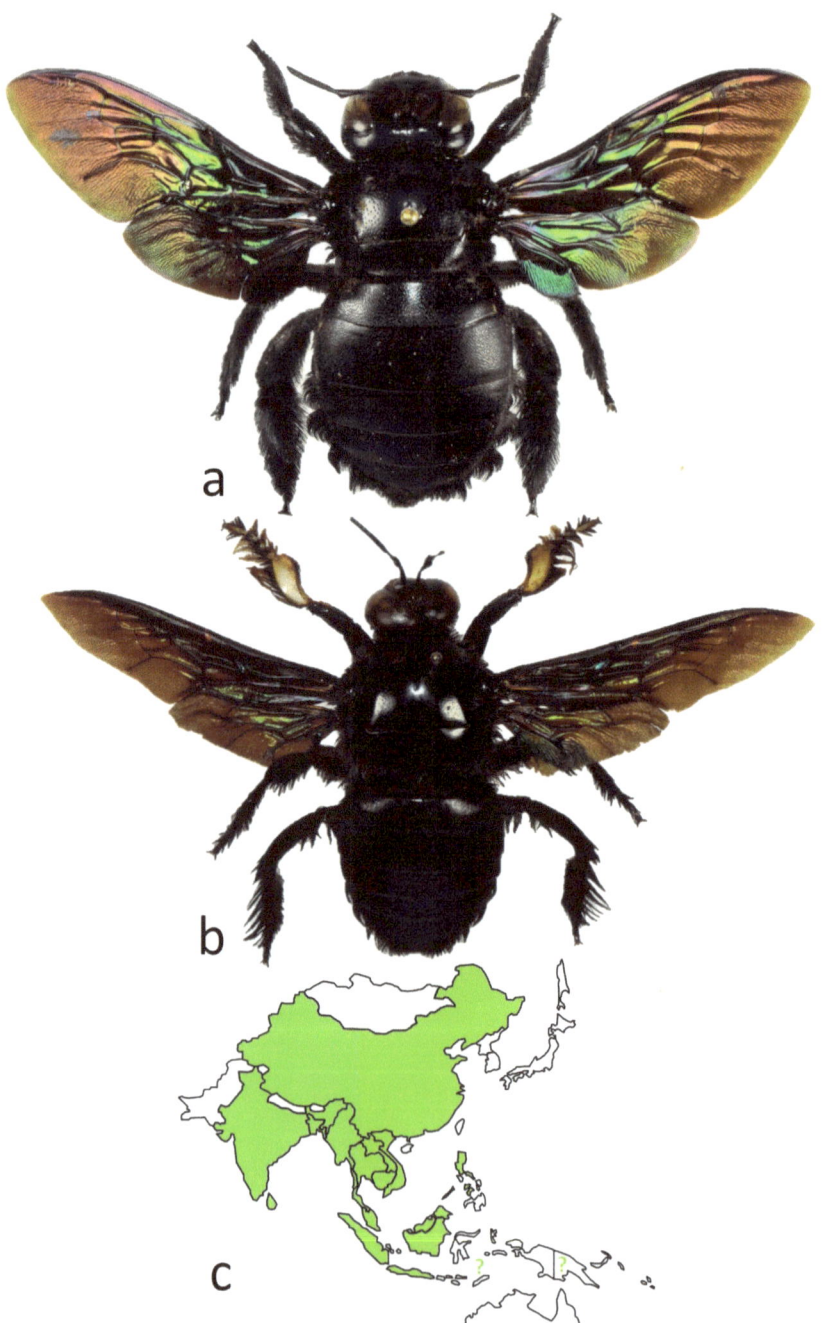

Figure 12: *Xylocopa latipes*

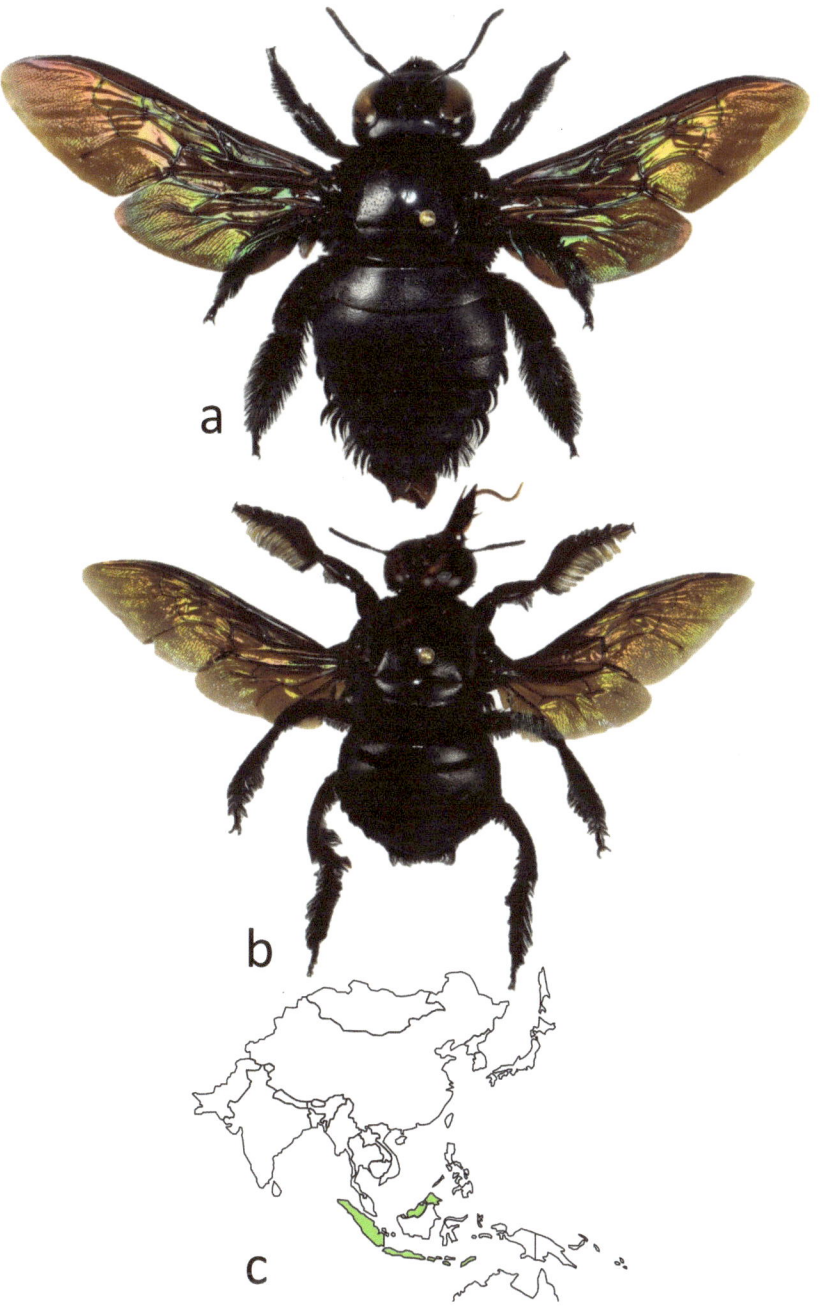

Figure 13: *Xylocopa perforator*

Giant Carpenter Bees

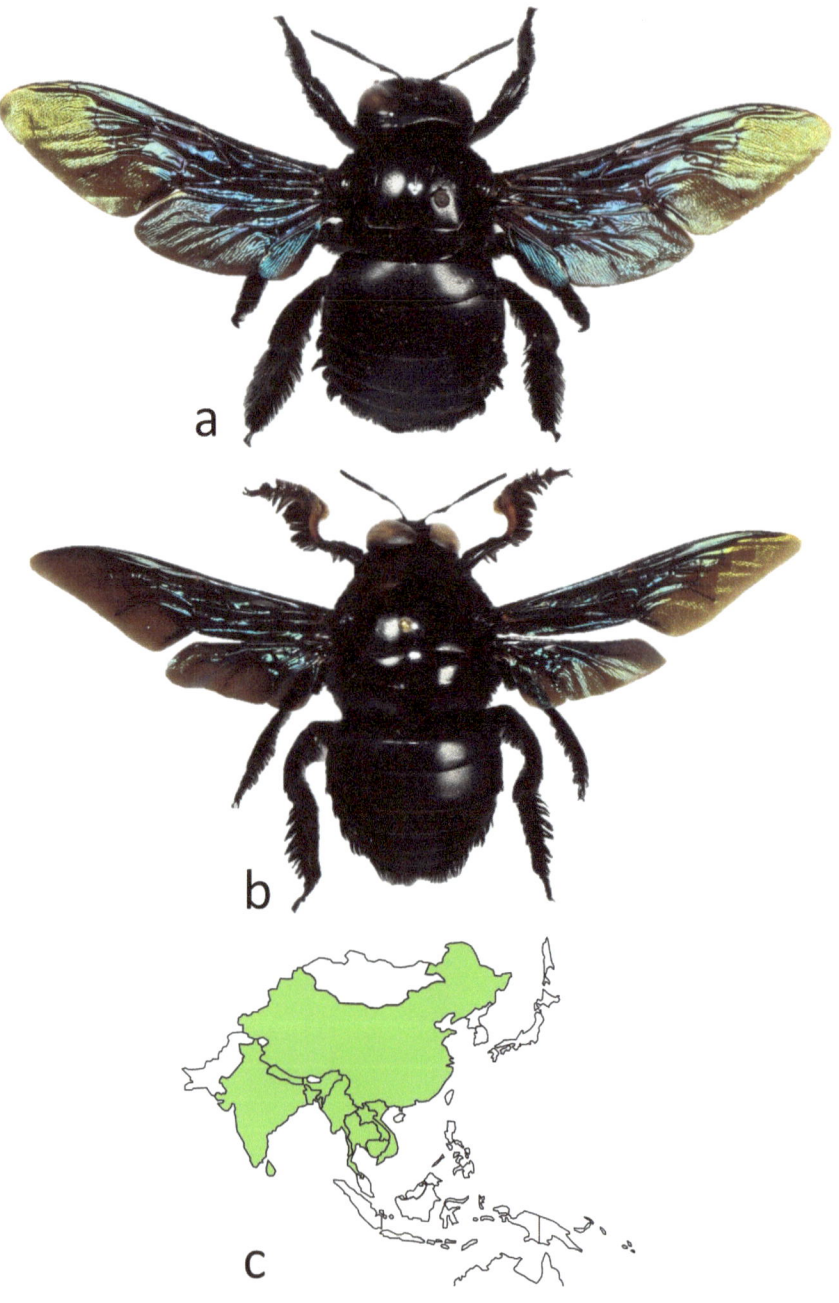

Figure 14: *Xylocopa tenuiscapa*

Figure 15: Species of *Xylocopa* subgenus *Biluna* from Southeast Asia and Indonesia

References

Ashmead, W. H. 1899. Classification of the bees, or the superfamily Apoidea. Transactions of the American Entomological Society 26:49-100.

Blanchard, E. 1844. Insectes recueillis à l'Himalaya, par Victor Jacquemont, pp. 13-31 in Jacquemont, Victor, Voyage dans l'Indie par Victor Jacquemont, pendant les années 1828 à 1832, vol. 4, Zoology. Paris: Didot Frères. 183 pp.

Cockerell, T. D. A. 1917. The carpenter bees of the Philippine Islands. Philippine Journal of Science 12(D):345-349

Cockerell, T. D. A. 1929. Descriptions and records of bees – CXIX. Annals and Magazine of Natural History 10(4):296-304.

DeGeer 1773. Mémoires pour servir à l'histoire des insectes, tome troisème. Stockholm: Pierre Hesselberg. xii + 950 pp. + 49 pls.

DeGeer, C. 1778. Mémoires pour servir à l'histoire des insectes, tome septième. Stockholm: Pierre Hesselberg. xii + 950 pp. + 49 pls.

Eardley, C. D. 1983. A taxonomic revision of the genus *Xylocopa* Latreille (Hymenoptera: Anthophoridae) in southern Africa. Entomology Memoir, Department of Agriculture, Republic of South Africa 58:1–67.

Eardley, C. D. 1987. Catalogue of Apoidea (Hymenoptera) in Africa south of the Sahara, Part I, The genus *Xylocopa* Latreille (Anthophoridae). Entomology Memoir, Department of Agiculture and Water Supply, Republic of South Africa 70:1-20.

Friese, H. 1911. Neue arten der bienengattung *Xylocopa* (Hym.). Deutsche Entomologische Zeitschrift 1911:685-687.

Friese, H. 1922. III Nachtrag zu "Bienen Afrikas." Zoologische Jahrbücher, Abteilung für Systematik, Geographie und Biologie der Tiere 46:1-42.

Hurd, P. D. and J. S. Moure. 1963. A classification of the large carpenter bees (Xylocopini) (Hymenoptera: Apoidea). University of California Publications in Entomology 29:vi + 365.

Keasar, T. 2010. Large carpenter bees as agricultural pollinators. Psyche 2010:1-7.

Lepeletier, A. L. M. 1841. Histoire naturelle des insects, Hyménoptères Suites à Buffon, tome 2. Paris: Roret. 680 pp.

LeVeque, N. 1928. Carpenter bees of the genus *Mesotrichia* obtained by the American Museum Congo Expedition, 1909-1915. American Museum Novitates 300:1-23.

Lieftinck, M. A. 1955. The carpenter-bees (*Xylocopa* Latr.) of the Lesser Sunda Islands and Tanimbar (Hymenoptera, Apoidea). Verhandlungen der Naturforschenden Gesellschaft Basel 66:5-32.

Maa, T. C. 1938. The Indian species of the genus *Xylocopa* Latr. (Hymenoptera). Records of the Indian Museum 40:265-329.

Maa T. 1940b. *Xylocopa* orientalia critica (Hymen.), III: Subgenus *Platynopoda* Westw. Lingnan Science Journal 19(4):565-575.

Mawdsley, J. R. 2015. An annotated checklist of the large carpenter bees, genus *Xylocopa* Latreille (Hymenoptera: Apidae), from the Philippine Islands. Oriental Insects 49(1-2):47-67.

Mawdsley, J. R. 2017. Taxonomy of the African large carpenter bees of the genus *Xylocopa* Latreille, 1802, subgenus *Xenoxylocopa* Hurd & Moure, 1963 (Hymenoptera, Apidae). Zookeys 655:131-139.

Mawdsley, J., J. Harrison, and H. Sithole. 2016. Natural history of a South African insect pollinator assemblage (Insecta: Coleoptera, Diptera, Hymenoptera, Lepidoptera): diagnostic notes, food web analysis and conservation recommendations, Journal of Natural History 50:2849-2879.

Michener, C. D. 2007. The bees of the world, second edition. Baltimore: Johns Hopkins University Press. 992 pp.

Radoszkowski, O. 1876. Compte-Rendu des Hyménoptères recueillis en Egypte et Abyssinie en 1873. Horae Societatis Entomologicae Rossicae 12:111-150.

Radoszkowski, O. 1881. Hyménoptères d'Angola. Jornal de sciencias mathematicas, physicas e naturaes 8(31):197-221.

Raju, A. J. S., and S. P. Rao. 2006. Nesting habits, floral resources and foraging ecology of alrge carpenter bees (*Xylocopa latipes* and *Xylocopa pubescens*) in India. Current Science 90(9):1210-1217.

Ritsema, C. 1876. Acht nieuwe oost-Indische *Xylocopa*-soorten. Tijdschrift voor Entomologie 19:177-185,

Ritsema, C. 1880. Description of three new exotic species of the hymenopterous genus *Xylocopa*. Notes from the Leyden Museum 2:220-224.

Sichel, J. 1867. Hymenoptera mellifera. Pp. 143-156 in Reise der Österreichischen fregatte Novara um die Erde in den Jahren 1857, 1858, 1859 unter den Befehlen des Commodore B. von Wüllerstorf-Urbair, Zoologischer Thiel, Zweiter Band. Vienna: Karl Gerold's Sohn. 104 pp. + 2 pls.

Smith, F. 1854. Catalogue of hymenopterous insects in the collection of the British Museum, part 2, Apidae. 465 pp. + 12 pls.

Smith, F. 1861. Catalogue of hymenopterous insects collected by Mr. A. R. Wallace in the islands of Ceram, Celebes, Ternate, and Gilolo. Journal of the Proceedings of the Linnaean Society of London 6:36-66.

Smith, F. 1874. Monograph of the genus *Xylocopa* Latr. Transactions of the Entomological Society of London 1874:247-302.

Vachal, J. 1900. Essai d'une revision synoptique des especes Europeennes et Africaines du *Xylocopa* Latr. (Hym.). Miscellanea Entomologica 8:106-108.

Vachal, J. 1922. Mellifères. Pp. 983-996 in Voyage de M. le baron Maurice de Rothschild en Éthiopie et en Afrique orientale anglaise (1904-1905): résultats scientifiques: animaux articulés, Deuxième Partie. Paris: Impremerie Nationale. 484-1015 pp.

Watmough, R. H. 1974. Biology and behaviour of Carpenter bees in southern Africa. Journal of the Entomological Society of Southern Africa 37(2):261-281.

Westwood, J. O. 1838. Description of a new genus of exotic bees. Transactions of the Entomological Society of London 2(2):112-113.

Westwood, J. O. 1840. The natural history of bees. The Naturalists' Library 6:17-31 + pls 1-30.

Wickler, W. 1968. Mimicry in plants and animals. New York: McGraw-Hill. 253 pp.

Wu, Y. 1982. A study of Chinese *Xylocopa* with description of new species *X. yunnanensis*. Zoological Research 3(2):193-200.

Glossary

Frons: The front area of the insect's head; in carpenter bees, it is the portion of the bee's head between its two compound eyes.

Integument: The hardened outer covering of the insect's body.

Mesonotum: The dorsal sclerite of the second thoracic segment in insects; in bees, it is the large flattened or dome-shaped sclerite on the dorsal surface of the mesosoma.

Mesosoma: The middle part of the bee's body, which is composed of the three thoracic segments and the first abdominal segment.

Metasoma: The hind part of the bee's body, which is composed of the remaining abdominal segments.

Metatibiae: The fourth segment of the hindmost (third) pair of legs in insects.

Ocellus (plural: **Ocelli**): The simple eyes on the bee's head (as opposed to the two large compound eyes).

Pubescence: Fine hairs on the surface of an insect.

Sclerite: A hardened body part of an insect.

Seta (plural: **Setae**): Larger, stout bristles on the surface of an insect.

Tergites: The dorsal sclerites of the insect's body; in bees, this term usually refers to the dorsal sclerites of the metasoma.

Vestiture: The pubescence, setae, or "hairs" covering the body of an insect.

Definitions of other entomological terms can be found at the website: https://en.wikipedia.org/wiki/Glossary_of_entomology_terms